KU-257-837

Symmetry in Molecules

Symmetry in Molecules

J. Michael Hollas
Lecturer in Chemistry and Chemical Physics
University of Reading

Chapman and Hall Ltd
11 New Fetter Lane, London EC4

First published 1972

Printed in Great Britain by William Clowes & Sons Ltd,
London, Colchester and Beccles

SBN 412 09570 X

Distributed in the USA
by Barnes and Noble Inc.

Contents

Preface

The application of group theory to molecules has far-reaching consequences in all branches of chemistry and it is becoming increasingly true that no one who calls himself a chemist or a chemical physicist can afford to be without a basic knowledge of this subject.

This book has its origins in undergraduate lecture courses which are given to students in Chemical Physics and Chemistry at Reading University. The material covered is intended to be understandable at undergraduate level although there is probably more information than can be covered in most undergraduate courses. The first five chapters of this book aim to present the subject in such a way that it is understandable to a first- or second-year undergraduate. The last two chapters are more advanced and deal with material which could be appreciated by third-year undergraduates and postgraduates.

Group theory was developed by mathematicians starting in the early nineteenth century and it was much later, in the 1920's and 1930's, that it was applied to atoms and molecules. Because of this historical development the subject has, in the past, often been taught in chemistry departments by starting with the theory of abstract mathematical groups and then applying this to molecular point groups. But most of us who are in chemistry departments find ourselves there, rather than in a mathematics department, because we find it easier to think in the particular rather than in the abstract. Consequently an audience of chemists or chemical physicists can easily be left behind in tackling the problems of molecular symmetry by starting with the theory of abstract groups.

It has been one of my aims in this book to keep the student in touch with chemistry by constantly giving examples. It is only at the end of Chapter 3 that some discussion of the theory of groups in general (as an abstract concept) is related to the theory of molecular point groups.

I am grateful to Professor I. M. Mills for encouraging me to write this book and to Dr. J. K. G. Watson, Dr. C. M. Woodman and Dr. T. Cvitaš who have been very helpful not only in reading and criticizing the manuscript but also in helping to clarify many points. Thanks are due also to Dr. S. N. Thakur for helping with proof-reading.

Introduction 1

1.1 General approach to molecular symmetry

If we were asked the question, 'which has the higher symmetry, a circle or a square?' most of us would instinctively judge correctly that the circle has the higher symmetry. But if the point were pressed further and we were asked to explain the reasons for the choice it might not be easy to put the instinctive feeling into words. It would not be too difficult to judge also that a rectangle has lower symmetry than either a circle or a square. In assessing the relative degrees of symmetry of an isosceles triangle and a parallelogram, however, some difficulty might be encountered. It is clear that both are fairly symmetrical but in different ways and it might not be possible to judge their relative symmetries in a meaningful way.

In considering the symmetry of molecules it is important to realise from the outset that the shape of the molecule under consideration must be known experimentally.* For example, in an assessment of the symmetry of the benzene molecule it is necessary to know that it has been shown experimentally that the carbon and hydrogen atoms each form regular hexagons with a common centre, that each carbon-hydrogen pair is on the same line radiating from the centre and that all the atoms are coplanar. Because of the regular hexagonal structure of benzene it is essential when considering its symmetry properties not to treat it as having, say, one of the Kekulé structures, shown in figure 1.1, or as being the enlongated hexagon we tend to draw and which even turns up in some textbooks! Less obvious examples where care has to be taken are those of molecules with a five-membered ring attached to another ring as in indane, whose molecular formula is often represented as in

* 'Interatomic Distances' published by the Chemical Society (London) is a good reference book for such data.

figure 1.2 whereas the true shape of the molecule is closer to that shown in figure 1.3. (The carbon atoms of the five-membered ring are probably not all coplanar, but that need not concern us here.) In cases of complex three-dimensional molecules, such as triethylene diamine, a representation of the

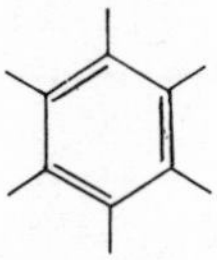

FIGURE 1.1
Kekulé structure of benzene

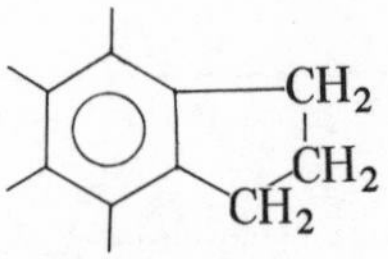

FIGURE 1.2
Misleading structure of indane

molecule such as that in figure 1.4 is often used but this does not illustrate the equivalence of all three $-CH_2-CH_2-$ chains nor, of course, the cage structure of the molecule which is shown in figure 1.5. Molecular models,

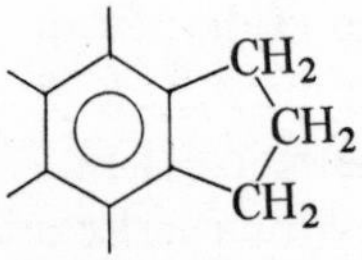

FIGURE 1.3
Realistic structure of indane

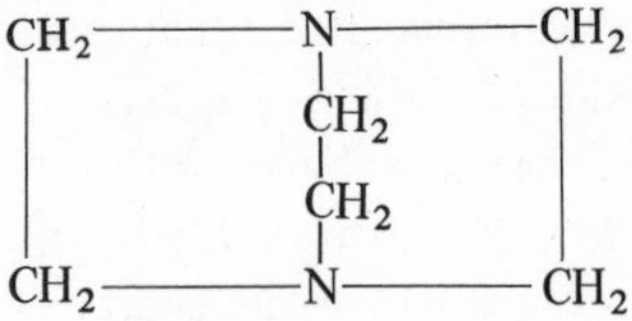

FIGURE 1.4
Misleading structure of triethylene diamine

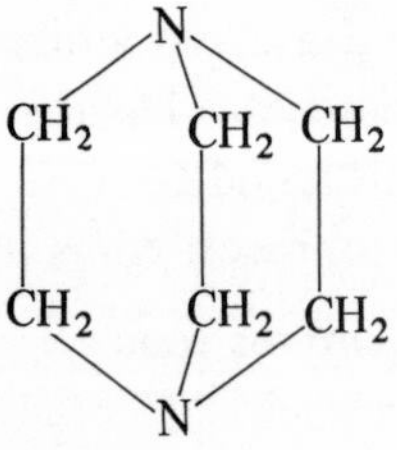

FIGURE 1.5
Realistic structure of triethylene diamine

which need not be very sophisticated, are a great help in assessing the symmetry properties of molecules whose structures are known.

Assuming that the molecular structures are known we can ask similar questions about their symmetry to those in the cases of geometrical figures. For example, we can consider whether ethylene (figure 1.6(a)) is more symmetrical than *trans*-difluoroethylene (figure 1.6(b)) and whether *cis*-difluoroethylene (figure 1.6(c)) is more symmetrical than the *trans* isomer.

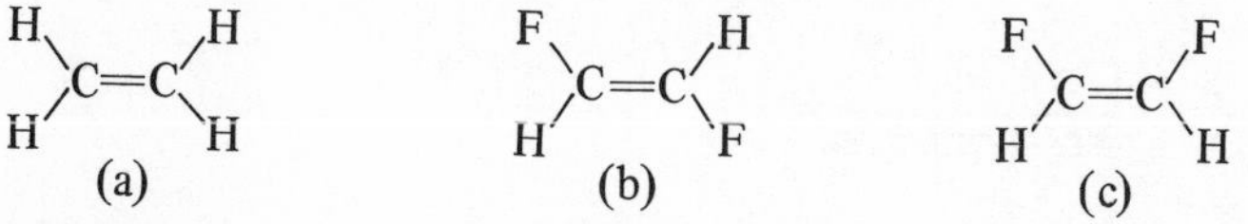

FIGURE 1.6
(a) Ethylene, (b) *trans*-difluoroethylene, (c) *cis*-difluoroethylene

It is fairly obvious that ethylene is of higher symmetry than either of the difluoroethylenes, but it is not at all obvious which of the difluoroethylenes has the higher symmetry. Indeed it looks as if it may not be possible, as in the case of the isosceles triangle and the parallelogram, to decide.

In considering symmetry in more detail it is clear that we have to put our assessments on a much more rigorous basis than we have done so far.

Before doing this it may be useful to get some idea of the kinds of problems which we are setting out to solve by molecular symmetry. One of these problems is that of selection rules for electronic transitions in atoms and molecules.

In the Bohr treatment of the hydrogen atom the electronic energy levels are given by the expression

$$E_n/hc = -R_{\mathrm{H}}/n^2 \tag{1.1}$$

where E_n is the total electronic energy, h is Planck's constant, c is the velocity of light, R_{H} is the Rydberg constant for hydrogen and n is a quantum number which can take any non-zero integral value, i.e. 1, 2, 3, 4 . . . These energy levels are illustrated on the left of figure 1.7. Electronic transitions between these levels are governed by *selection rules* which tell us whether transitions between the levels are *allowed* or *forbidden.* In this case the selection rule is the rather trivial one that Δn is unrestricted. When the effects of elliptical orbits and relativity were included by Sommerfeld it was found that each level is split into n sub-levels, which can be described by the quantum number l where $l = 0, 1, 2, \ldots (n - 1)$. These sub-levels are shown on the right of figure 1.7. The selection rule for transitions involving the sub-levels is $\Delta l = \pm 1$. In the Sommerfeld treatment of the hydrogen atom the selection rules are described in terms of the quantum numbers n and l only. In the later quantum mechanical treatment the selection rules are described in terms of four quantum numbers n, l, s, and j.

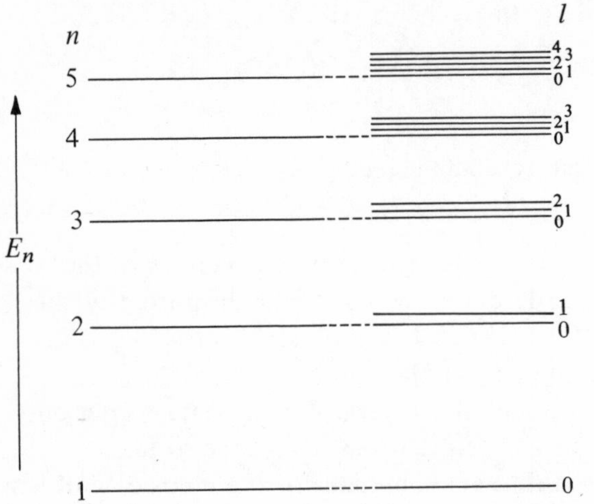

FIGURE 1.7
Schematic energy level diagram of atomic hydrogen

A similar state of affairs exists for polyelectronic atoms in that selection rules can be described completely in terms of quantum numbers. In diatomic molecules, however, quantum numbers alone are not sufficient to describe the selection rules: simple symmetry properties of the electronic wave function have to be used as well. For example, in a homonuclear diatomic molecule such as nitrogen, transitions are not allowed between electronic states whose wave functions are either both symmetrical or both unsymmetrical about the centre of the molecule. In linear polyatomic molecules the situation is very similar to that in diatomics, but in non-linear polyatomics there is only one quantum number, that associated with electron spin, with which to describe selection rules for electronic transitions: otherwise symmetry properties alone have to be used.

At this stage one might reasonably wonder why the ways of describing electronic selection rules are apparently so different in atoms from those in non-linear polyatomic molecules. The answer is that they need not be. The selection rules in all atoms and molecules can be expressed in terms of symmetry properties. However, there is something of a paradox in that as we go from the simplest atom to the most complex polyatomic molecule the symmetry properties become simpler, rather than more complex. Consequently the quantum number description in atoms is simpler than the symmetry property description.

Selection rules governing transitions between vibrational energy levels in molecules are also described in terms of symmetry properties, although in all cases there are vibrational quantum numbers, v_i, which are also involved.

Other important uses of symmetry properties are in molecular orbital calculations and in correlating molecular orbitals between a reactant and a product in order to make a prediction about the product which will be formed.

Before we start to systematize molecular symmetry it is as well to realize that in spite of the fact that in general we shall get specific answers to many problems, there are cases of molecules in which, at first sight, the answers may not be quite so specific. For example, the ammonia molecule is known to be pyramidal in its ground electronic state and to be planar in some of its excited electronic states. How do we classify the electronic states: in terms of a planar or a pyramidal configuration? The question of isotopic substitution on symmetry is also an interesting one. How do we classify, for example, monodeutero-benzene: in the same way as benzene or, say, fluorobenzene? Both these types of problem are important and we shall return to them in Chapters 5 and 7.

1.2 The electromagnetic spectrum and the Born-Oppenheimer approximation

Some of the most important applications of symmetry in molecules follow from the classification of the electronic, vibrational and rotational wave functions, according to their symmetry properties. The approximation involved in factorization of the total wave function of a molecule into the electronic, vibrational and rotational parts is known as the Born-Oppenheimer approximation. The validity of this approximation can be demonstrated by considering the regions in the electromagnetic spectrum in which the three processes of electronic, vibrational and rotational excitation occur.

When a molecule is irradiated with electromagnetic radiation of wavenumber $\tilde{\nu}$ the effect on the molecule may be more or less drastic according to the wavenumber of the radiation, since the wavenumber is proportional to the energy as given by the equation

$$E = hc\tilde{\nu} \tag{1.2}$$

where E is the energy, h is Planck's constant and c is the velocity of light.

The electromagnetic spectrum is shown schematically in figure 1.8. The units used are those of cm^{-1}, a unit of wavenumber, and also nanometres,† a unit of wavelength (λ) which is then related to $\tilde{\nu}$ by

$$10^7/\lambda = \tilde{\nu} \tag{1.3}$$

The division into regions such as the ultraviolet, visible and infra-red is also shown. However, it is important to realize that the division is really an artificial one which is useful mainly because of the different experimental techniques required in the various regions. The energy of the radiation increases from the radio wave to the γ-ray region. This is apparent when we remember how harmful γ-rays are to the human body compared with harmless radio waves.

One of the effects on a molecule (or atom) which requires the most energy is the removal of an electron from one of the orbitals in an ionization process. This occurs typically in the vacuum ultraviolet region of the spectrum. Less energy is required in general to promote an electron from one orbital to another. This occurs usually with visible or ultra-violet light although there is some overlap into the infra-red and vacuum ultra-violet regions. It requires less energy for a vibrational transition which occurs in the presence of infra-red radiation. Still less energy is required for a rotational transition which may occur with far infra-red or microwave radiation.

† 1 nm = 10 Å.

From figure 1.8 we see that a typical electronic excitation might occur at a wavenumber of about 30 000 cm^{-1} or a frequency ν ($= c\bar{\nu}$)† of 9×10^{14} s^{-1}. A typical vibrational excitation has a wavenumber of about 1 000 cm^{-1} and a frequency of 3×10^{13} s^{-1}, whereas a typical rotational excitation has a wavenumber of about 10 cm^{-1} and a frequency of 3×10^{11} s^{-1} By comparing the frequencies of these processes we can see that an

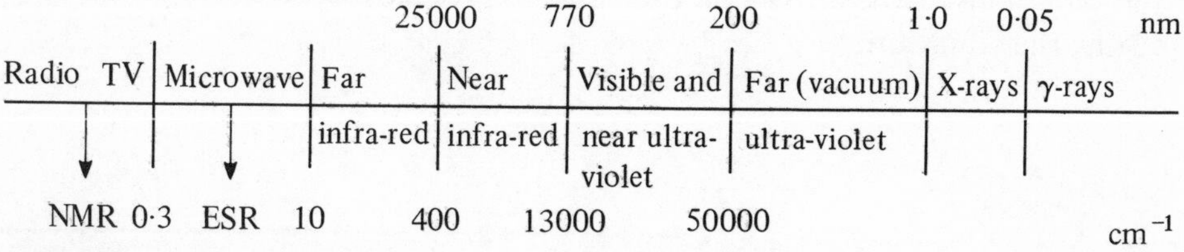

FIGURE 1.8
Schematic representation of the electromagnetic spectrum

electronic transition is about thirty times faster than a vibrational transition, which is about a hundred times faster than a rotational transition. These large differences in the time scale of the three processes mean that each can be considered, for many purposes, to be independent of the other two. The approximation involved in treating the processes separately is the Born-Oppenheimer approximation. Important results of the approximation are:

(*a*) the total wave function (excluding nuclear spin) of a molecule can be expressed as the product of the electronic, vibrational and rotational wave functions

$$\psi = \psi_e \times \psi_v \times \psi_r \tag{1.4}$$

(*b*) the total energy of a molecule can be expressed as the sum of electronic, vibrational and rotational energies

$$E = E_e + E_v + E_r \tag{1.5}$$

† It is unfortunate that the symbol ν is used commonly for frequency *and* wavenumber. In this book it will be used for frequency only.

In later chapters we shall be closely concerned with the symmetry properties of ψ_e and ψ_v. Except in a few special cases, such as isotopic substitution, neither ψ_e nor ψ_v can have higher symmetry than the nuclear configuration of the molecule. They often have less symmetry as, for example, an electronic wave function of the water molecule illustrated in figure 1.9 or a vibrational wave function of the acetylene molecule represented by the nuclear motions illustrated in figure 1.10: the lengths and directions of the arrows on the nuclei represent their relative motions during the vibration.

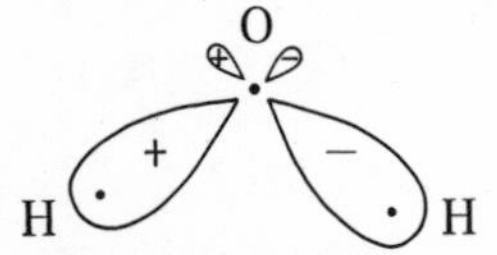

FIGURE 1.9
An electronic wave function of the water molecule

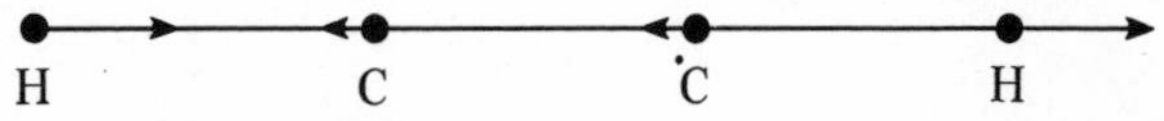

FIGURE 1.10
A vibration of acetylene

Molecules are classified according to the symmetry of the stable, equilibrium nuclear configuration and it is this classification with which we shall first be concerned. Then we shall go on to see how to deal with situations in which some of this symmetry is lost, as in the examples given above of electronic and vibrational wave functions.

1.3 Normal vibrations and normal co-ordinates

In discussing vibrational wave functions it will be necessary to know what is meant by a *normal vibration* and a *normal co-ordinate.*

A diatomic molecule can vibrate in only one way. To a fairly good approximation in many cases this vibration, especially if it is of low amplitude, is a simple harmonic motion about the equilibrium position. At this position the internuclear distance is r_e, illustrated in figure 1.11. The potential energy $V(r)$ of the molecule has a minimum when $r = r_e$ and, in the simple harmonic oscillator approximation, it is given by

$$V(r) = \tfrac{1}{2}k(r - r_e)^2 \tag{1.6}$$

where k is the force constant and $(r - r_e)$ is the displacement of the internuclear distance from its equilibrium value. The broken parabola in figure 1.12 is the potential energy curve given by equation 1.6. However, vibrational motion deviates from that of a simple harmonic oscillator for two main

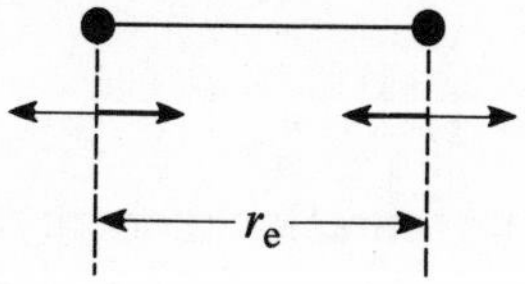

FIGURE 1.11
Vibration about the equilibrium position in a diatomic molecule

reasons: (*i*) as the nuclei come close together at small values of r, nuclear repulsion causes the potential energy to increase at a greater rate than in the simple harmonic oscillator; (*ii*) as the nuclei move further apart there is a weakening of the bond, which eventually leads to dissociation. The model which corresponds to this type of potential energy curve, shown also in figure 1.12, is that of the anharmonic oscillator. The dissociation energy, D_e, is also indicated in the figure.

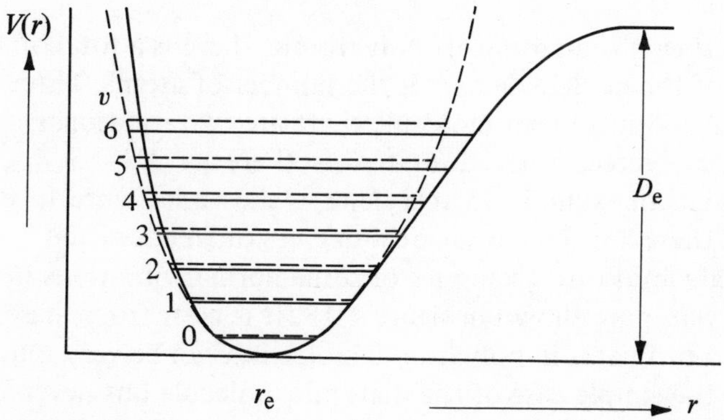

FIGURE 1.12
Potential energy curves for a harmonic oscillator (broken line) and an anharmonic oscillator (solid line) model of a diatomic molecule.

In an anharmonic oscillator the potential energy is expressed as a power series in $(r - r_e)$ in which the cubic and quartic terms are typically important.

Whether or not the vibration is simple harmonic in character, the mode of vibration, of which there is only one in a diatomic molecule, is called a normal mode and the co-ordinate $(r - r_e)$ is the normal co-ordinate.

The vibrational energy E_v is quantized and is given for the simple harmonic oscillator by

$$E_v = hc\tilde{\nu}(v + \tfrac{1}{2}) \tag{1.7}$$

where E_v is in energy units. Alternatively the vibrational energy levels can be expressed by

$$G_v = \omega_e(v + \tfrac{1}{2}) \tag{1.8}$$

where G_v is in wavenumber units and ω_e is the vibration wavenumber. In both equations v is the vibrational quantum number. The resulting equally-spaced energy levels are shown by the broken lines in figure 1.12.

In an anharmonic oscillator, equation 1.8 is modified to

$$G_v = \omega_e(v + \tfrac{1}{2}) + x_e(v + \tfrac{1}{2})^2 + y_e(v + \tfrac{1}{2})^3 + \cdots \tag{1.9}$$

where x_e, y_e . . . are anharmonicity constants of rapidly decreasing magnitude with higher powers of $(v + \frac{1}{2})$.† The usual effect of anharmonicity on the vibrational energy levels is to make them close up as v increases. This is shown in figure 1.12.

In any molecule, whether it is diatomic or polyatomic, there is a total of $3N$ degrees of freedom of the nuclei where N is the number of atoms. Three of these are translational and in a linear molecule there are two rotational, leaving $3N - 5$ vibrational degrees of freedom; that is, there are $3N - 5$ normal vibrations of which an example, in acetylene, is shown in figure 1.10. In non-linear molecules there are three rotational degrees of freedom and therefore $3N - 6$ normal vibrations. Examples of some normal vibrations in ammonia, benzene and water are shown in figure 1.13. It is clear from these examples that normal co-ordinates in polyatomic molecules can be very much more complicated than the simple case of the diatomic molecule but nevertheless we can still use the concept of a potential energy curve for each

† $x_e, y_e, \ldots$ are usually replaced by $-\omega_e x_e, \omega_e y_e, \ldots$ in diatomic molecules.

normal mode of vibration. For each curve, potential energy in that degree of freedom is plotted against the normal co-ordinate. Each curve will usually resemble qualitatively the solid curve in figure 1.12.

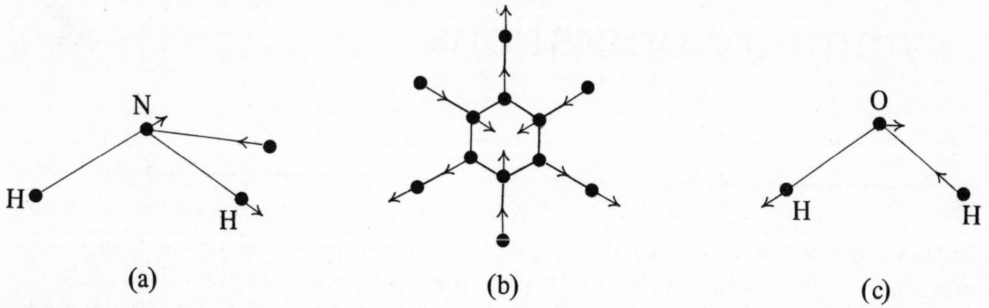

FIGURE 1.13
Examples of normal vibrations in (a) ammonia, (b) benzene and (c) water

Symmetry elements and symmetry operations

2

In this book we shall be considering the symmetry properties of so-called *free molecules* only. In the present context this description implies that each molecule is not influenced from the point of view of its geometry by interaction with any neighbouring molecules. Such conditions obtain when the molecules are in the vapour phase at low pressures. In the rather special case of a molecular crystalline solid, the regular arrangement of molecules within the crystal lattice can usefully be treated by considering the symmetry properties of the arrangement, but we shall not be concerned here with those additional symmetry properties. However, it is well to point out that crystallographers traditionally use a different notation for symmetry properties than that usually used for the free molecule. They use the Hermann-Maugin (H-M) system as opposed to the free molecule Schoenflies system. In this chapter the H-M equivalent of each of the Schoenflies symbols introduced is given and a correlation table for the two sets of symbols given in section 2.6.

The symmetry of a free molecule can be described completely in terms of *symmetry elements* of which there are only five kinds and one of these is trivial. We shall now consider these symmetry elements in detail.

2.1 Axis of symmetry—C_n

A molecule having a C_n axis of symmetry can be rotated by $2\pi/n$ radians about the axis and the configuration will remain unchanged i.e. the final configuration is indistinguishable, with respect to external axes, from the initial one. n is always an integer. Figure 2.1 shows some examples of molecules having a C_n axis of symmetry. Water (figure 2.1(a)) has a C_2 axis since rotation by π radians about this axis simply exchanges the positions of equivalent hydrogen atoms. Similarly ammonia (figure 2.1(b)), which is pyramidal, has a C_3 axis, benzene (figure 2.1(c)) has a C_6 axis perpendicular to the plane of the molecule, the $[ICl_4]^-$ ion (figure 2.1(d)) is planar and has a C_4 axis, and HCN (figure 2.1(e)), like all linear molecules, has a C_∞ axis: rotation through any angle about the C_∞ axis leaves the configuration unchanged.

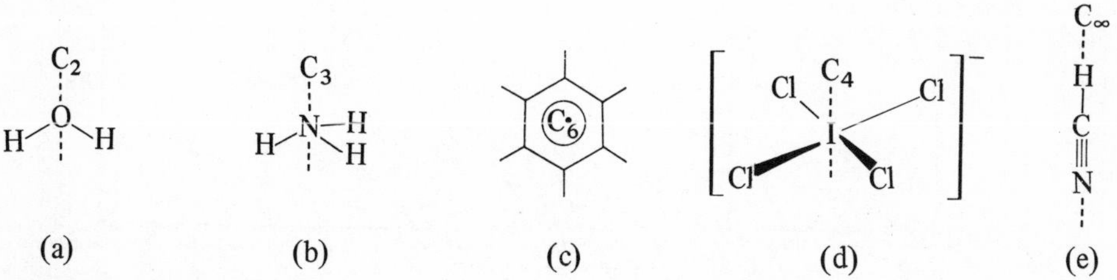

FIGURE 2.1
Illustration of some C_n axes

All symbols used to represent symmetry elements are also used to represent *symmetry operations.* For example the symbol C_n is a label not only for an n-fold axis of symmetry, but also for the operation of carrying out a *clockwise*† rotation of the molecule by $2\pi/n$ about the C_n axis.

The H-M symbol for an n-fold axis of symmetry is n, e.g. $C_2 \equiv 2$.

2.2 Plane of symmetry—σ

A molecule having a plane of symmetry σ is unchanged in configuration if all the atoms are reflected across the plane. If, for example, the plane of symmetry is the xy-plane then changing all the positions of the atoms from z to $-$z will not change the configuration.

Figure 2.2 shows some examples of molecules having a plane of symmetry. Difluoromethane (figure 2.2(a)) has two mutually perpendicular σ_v planes. Both planes contain the C_2 axis. The subscript 'v' on the labels of these planes stands for 'vertical' and means in general that if the C_n axis with the largest value of n in the molecule is regarded as vertical then these planes are also vertical. In the case of difluoromethane the C_2 axis is regarded as vertical

† Some authors define this rotation as *anti-clockwise.*

whereas in the planar molecule boron trifluoride (figure 2.2(b)), if the C_3 axis perpendicular to the plane of the molecule is vertical, then the plane of the molecule is horizontal and labelled σ_h. In aniline (figure 2.2(c)), in which

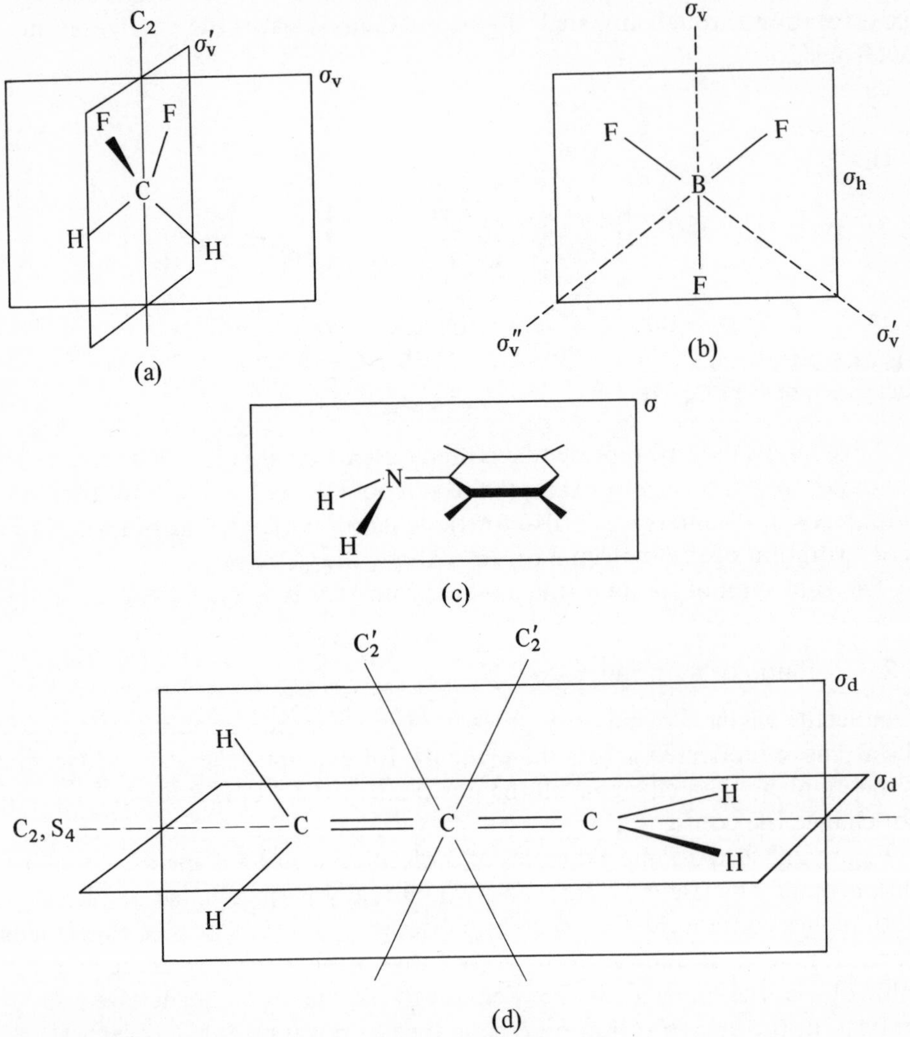

FIGURE 2.2
Illustration of some planes of symmetry

the hydrogen atoms of the amino group are not coplanar with the rest of the molecule, the plane of symmetry is simply labelled σ since the molecule has no axis of symmetry to define the vertical direction. Allene (figure 2.2(d)), in which the two CH_2 groups lie in planes perpendicular to each other, has two planes of symmetry which bisect the angles between two C_2' axes: these planes are labelled σ_d in which 'd' stands for 'diagonal'.

The symbol σ also represents the symmetry operation of reflecting the atoms across the plane of symmetry.

The H-M symbol for a plane of symmetry is m, with no distinction between the various types of plane.

2.3 Centre of symmetry—i

If a molecule has a centre of symmetry i, reflection of the positions of all the atoms through this centre, that is changing all their positions from (x,y,z) to (– x,– y,– z), will leave the configuration of the molecule unchanged. This process is called *inversion.*

Some examples of molecules having a centre of symmetry are illustrated in figure 2.3; these are the ferricyanide ion (figure 2.3(a)) which has a regular octahedral configuration, naphthalene (figure 2.3(b)), and *trans*-difluoroethylene (figure 2.3(c)). (Note that *cis*-difluoroethylene (figure 1.6(c)) does not have a centre of symmetry.)

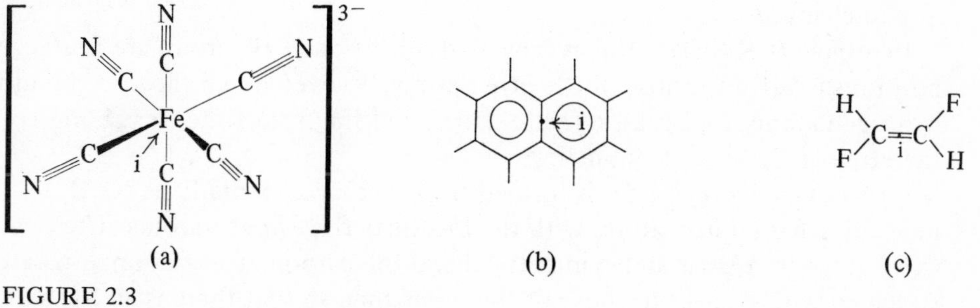

FIGURE 2.3
Illustration of some centres of symmetry

The symbol i also represents the symmetry operation of changing the positions of all the atoms in the molecule from (x,y,z) to (– x,– y,– z).

The H-M symbol for a centre of symmetry is $\bar{1}$. This symbol will be explained in section 2.4.

2.4 Rotation-reflection axis of symmetry—S_n

In a molecule having an n-fold rotation-reflection axis S_n, rotation by $2\pi/n$ radians about the axis, followed by reflection across a plane perpendicular to the axis and passing through the centre of the molecule, will leave the configuration unchanged. It is worth noting that the plane of reflection need not be a plane of symmetry.

The S_n symmetry element is often referred to as an *improper rotation axis* whereas C_n is a *proper rotation axis.*

It can easily be seen that $S_1 = \sigma$ and, for example in *trans*-difluoroethylene (figure 2.3(c)), that $S_2 = i$.

The planar BF_3 molecule (figure 2.2(b)) has an S_3 axis perpendicular to the σ_h plane. It is easy to see that one difference between a pyramidal configuration of an XY_3 molecule like ammonia (figure 2.1(b)) and a planar configuration like BF_3 is the absence or presence of an S_3 axis of symmetry. Allene (figure 2.2(d)) has an S_4 axis, which is the line of intersection of the two σ_d planes. Looking down the S_4 axis (figure 2.4(a)) it can be seen that a rotation by $2\pi/4$ radians, followed by a reflection across a plane perpendicular to this axis and passing through the centre of the molecule, leaves the configuration unchanged.

In ethane ($CH_3.CH_3$) the hydrogen atoms of one CH_3 group are in staggered positions relative to those of the other group. Viewed down the C—C bond the hydrogen atoms appear as in figure.2.4(b) and the line of the C—C bond is therefore an S_6 axis of symmetry.

Ferrocene (figure 2.4(c)) is formed from two cyclopentadiene (C_5H_6) molecules and an iron atom, with the loss of two hydrogen atoms. The two C_5H_5 rings are planar and symmetrical and the carbon atoms of one ring are staggered with respect to those of the other ring, so that there is an S_{10} axis as shown.

The symbol S_n is used also to represent the symmetry operation of rotation through $2\pi/n$ radians, followed by reflection across a plane perpendicular to the S_n axis and passing through the centre of the molecule.

The H-M system differs from the Schoenflies system in that it uses the rotation-inversion instead of the rotation-reflection symmetry element. If a molecule has an n-fold rotation-inversion axis of symmetry, rotation by $2\pi/n$ radians followed by inversion through the centre of the molecule (which need not be a centre of symmetry) leaves the configuration unchanged. The H-M symbol for this symmetry element is $\bar{n}$.

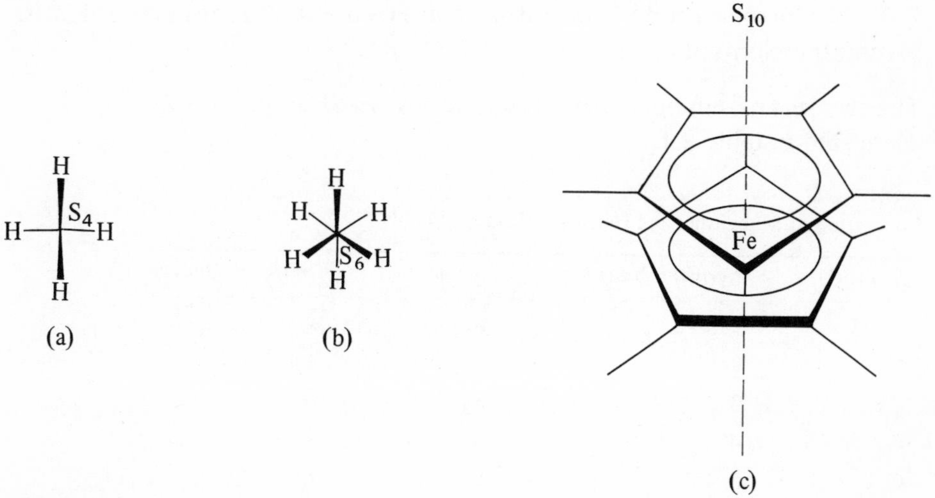

FIGURE 2.4
Illustration of some S_n axes

It can be seen using the examples of boron trifluoride (figure 2.2(b)) that S_3 implies $\bar{6}$, of allene (figure 2.4(a)) that S_4 implies $\bar{4}$, of ethane (figure 2.4(b)) that S_6 implies $\bar{3}$, and of ferrocene (figure 2.4(c)) that S_{10} implies $\bar{5}$. In general S_{2n+1} implies $\overline{4n+2}$ and S_{2n} implies $\bar{n}$ if n is odd and $\overline{2n}$ if n is even.

2.5 Identity element of symmetry—I

All molecules possess the identity element of symmetry I which is equivalent to C_1, that is, rotation of the molecule through 2π radians leaves the configuration unchanged. This symmetry element may seem trivial at this stage, but its use will become apparent in section 2.7.

The symbol I applies also to the symmetry operation which leaves the molecule unchanged.

In the older literature the symbol E has often been used instead of I but this use of E clashes with its use in another context (see chapter 4).

The H-M symbol 1 is equivalent to I.

2.6 Correlation of Schoenflies and Hermann-Maugin symbols for symmetry elements

The two sets of symbols introduced and discussed in sections 2.1 to 2.5 are correlated in table 2.1.

TABLE 2.1
Correlation of Schoenflies and Hermann-Maugin symbols

Schoenflies	*Hermann-Maugin*	*Schoenflies*	*Hermann-Maugin*
$I = C_1$	1	$S_1 = \sigma$	$\bar{2} \equiv m$
C_2	2	$S_2 = i$	$\bar{1}$
C_3	3	S_3	$\bar{6}$
C_4	4	S_4	$\bar{4}$
C_5	5	S_5	$\overline{10}$
C_6	6	S_6	$\bar{3}$
⋮	⋮	S_7	$\overline{14}$
C_n	n	S_8	$\bar{8}$
$\sigma_{v,h,d}$	m	⋮	⋮
i	$\bar{1}$	S_{2n+1}	$\overline{4n+2}$
		S_{2n}	$\bar{n}$ (n even) $\overline{2n}$ (n odd)

2.7 Multiplication of symmetry operations and elements

If we wish to carry out two symmetry operations A and B in turn then this multiple operation is written B x A: that is we carry out operation A first and then operation B. In the case of difluoromethane (figure 2.2(a)), the result of performing a C_2 followed by a σ_v operation is shown in figure 2.5 to be equivalent to performing the single operation σ_v' and we can express this equality as

$$\sigma_v \times C_2 = \sigma_v' \tag{2.1}$$

When σ_v and C_2 represent symmetry elements rather than symmetry operations, equation 2.1 tells us that if a molecule has a C_2 axis and one σ_v plane it must necessarily have a second σ_v plane. The symmetry elements C_2 and σ_v are said to *generate* the element σ_v'.

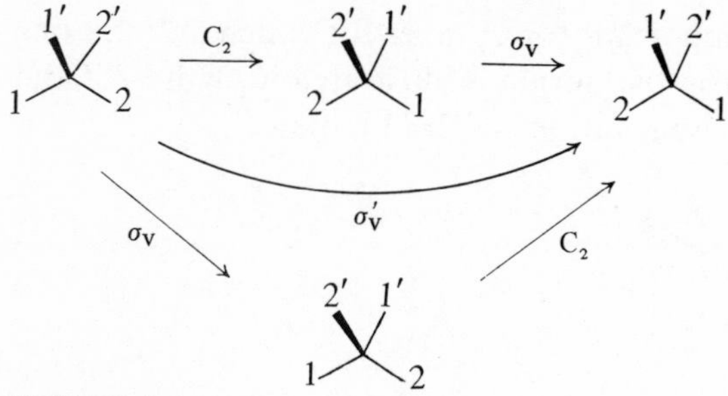

FIGURE 2.5
Demonstration that, in difluoromethane, $\sigma_v \times C_2 = C_2 \times \sigma_v = \sigma_v'$

In figure 2.5 it is demonstrated also that

$$\sigma_v \times C_2 = C_2 \times \sigma_v \tag{2.2}$$

If in general for two symmetry operations A and B, A x B = B x A then B and A are said to *commute* or to be *commutative*. If A x B ≠ B x A then A and B *do not commute* or are *non-commutative*. An example of a pair of

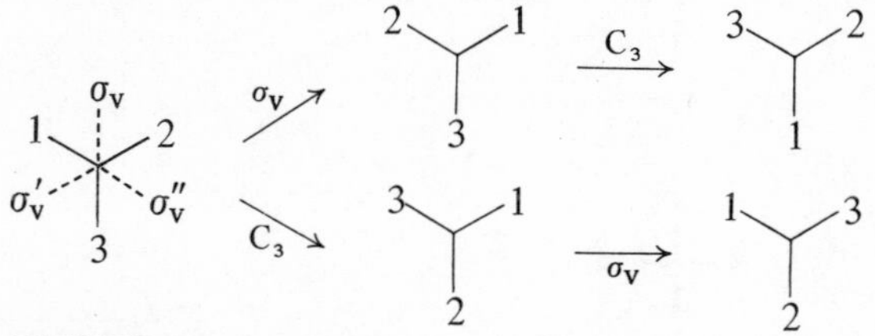

FIGURE 2.6
Demonstration that, in boron trifluoride, $C_3 \times \sigma_v \neq \sigma_v \times C_3$

operations which do not commute is the pair C_3 and σ_v in boron trifluoride (figure 2.2(b)). It is shown in figure 2.6 that

$$C_3 \times \sigma_v \neq \sigma_v \times C_3 \tag{2.3}$$

Symmetry operations can be carried out more than once and this is symbolized by raising the operation to the appropriate power. For example C_2^2 implies two clockwise rotations of $2\pi/2$ radians about the C_2 axis. The result is a rotation by 2π and therefore

$$C_2^2 = I \tag{2.4}$$

Similarly

$$\sigma_{h,v,d}^2 = i^2 = I \tag{2.5}$$

The direction of the rotation is important for C_n axes for which $n > 2$. For example the operation C_3^2 in boron trifluoride is illustrated in figure 2.7 and is not the same as two anticlockwise rotations of $2\pi/3$ radians.

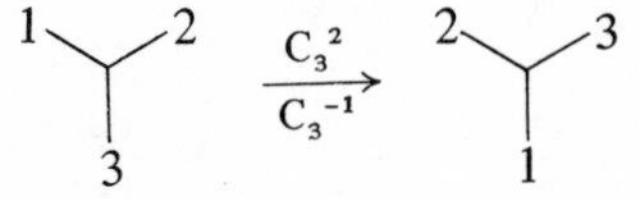

FIGURE 2.7
The C_3^2 (or C_3^{-1}) symmetry operation in boron trifluoride

The operation S_n may also be raised to any power. The example of allene (figure 2.8) shows that

$$S_4^2 = C_2 \tag{2.6}$$

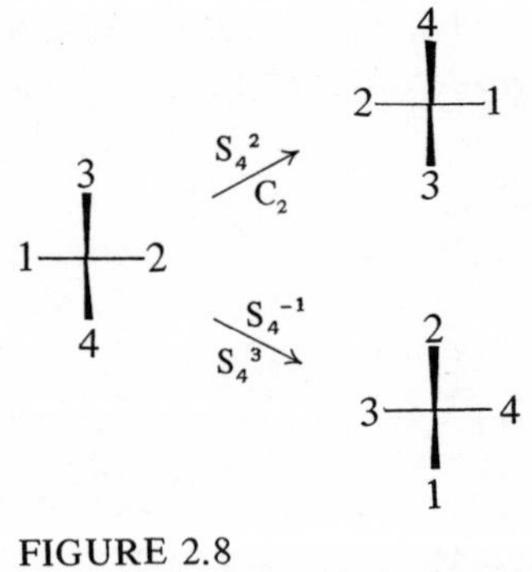

FIGURE 2.8
S_4^2 and S_4^3 symmetry operations in allene

In general it is obvious that

$$S_n^n = I \text{ if } n \text{ is even and } \sigma_h \text{ if } n \text{ is odd} \tag{2.7}$$

To every symmetry operation A there corresponds an *inverse operation* A^{-1}, which reverses the effect of A. It is clear that the operations σ and i are equal to their inverses

$$\sigma^{-1} = \sigma; \quad i^{-1} = i \tag{2.8}$$

and that

$$C_2^{-1} = C_2 \tag{2.9}$$

where C_2^{-1} implies an anticlockwise rotation by π radians about the C_2 axis. However for $n > 2$ the C_n operation is not equal to its inverse. This is illustrated for boron trifluoride in figure 2.7 which shows that

$$C_3^{-1} = C_3^2 \tag{2.10}$$

and, in general,

$$C_n^{-1} = C_n^{n-1} \tag{2.11}$$

The inverse operation S_n^{-1} implies an anticlockwise rotation by $2\pi/n$ radians, followed by a reflection across a plane perpendicular to the S_n axis and passing through the centre of the molecule. Since the reflection is its own inverse, it is only the inverse of the rotation which has to be considered. The S_4^{-1} operation is illustrated in figure 2.8 which demonstrates for the case of allene that

$$S_4^{-1} = S_4^3 \tag{2.12}$$

In general we have

$$S_n^{-1} = S_n^{2n-1} \text{ always and } S_n^{n-1} \text{ if } n \text{ is even} \tag{2.13}$$

Point groups

3

In any molecule at least one point remains unchanged no matter how many symmetry operations are carried out. An example of such a point is the centre of the benzene molecule (figure 2.1(c)). It is for this reason that the complete set of symmetry elements in any particular molecule is called a *point group*. A point group should be distinguished from a *space group* which describes a set of symmetry elements which includes translational elements. Such elements are relevant, not to the free molecule, but to the regular arrangements of molecules found in crystals.

A point group is a particular example of groups in general which form the basis of *group theory*. Group theory was originally developed by mathematicians and then applied later to the particular case of molecular symmetry. It is a consequence of this historical development that the subject of molecular symmetry is often introduced by a mathematical treatment of general groups before applying the results specifically to molecular point groups. It may be for this reason that understanding of molecular symmetry is not as widespread as it should be.

In section 3.11 the relation between general group theory and the theory of molecular point groups will be discussed briefly.

If all the symmetry elements present in known molecules were listed it would soon become apparent that the possible combinations of elements is limited, that is the number of point groups is limited. It follows that many different molecules may belong to the same point group; for example water (figure 2.1(a)) and difluoromethane (figure 2.2(a)) must belong to the same point group since they each possess only the symmetry elements I, C_2, σ_v, σ_v'. Further, although each point group is given a label e.g. C_{2v} for the combination of elements I, C_2, σ_v, σ_v', it is useful to collect together point groups which have certain types of elements in common. For example, the symbol $\mathbf{C}_{nv}$ represents the point groups having the combination of elements I, C_n, $n\sigma_v$. It is under the headings of such general point group symbols that point groups will now be discussed.

In listing the symmetry elements of the various point groups the identity element I, which necessarily is present in all point groups, will be omitted.

3.1 C_n point groups

A C_n point group contains the symmetry element C_n.

A C_n point group also contains the elements generated from C_n by raising it to the powers 2, 3, 4 . . . $(n-1)$† but, at this stage, when our primary concern is to be able to assign any molecule to a point group, these additional elements need not be considered further: their importance will become apparent in section 3.13.

(*i*) C_1. This point group contains only the $C_1 (= I)$ element, that is a rotation by 2π radians leaves the configuration unchanged. This is the lowest symmetry that a molecule can have: an example is the substituted methane CHFClBr illustrated in figure 3.1.

(*ii*) C_2. The only symmetry element, apart from I, in this point group is C_2. There are not many molecules belonging to this point group but hydrogen peroxide is a well-known example. The angle between the two O—O—H planes (figure 3.2) is about 111° so the molecule has only a C_2 symmetry axis. F_2O_2 is a similar example.

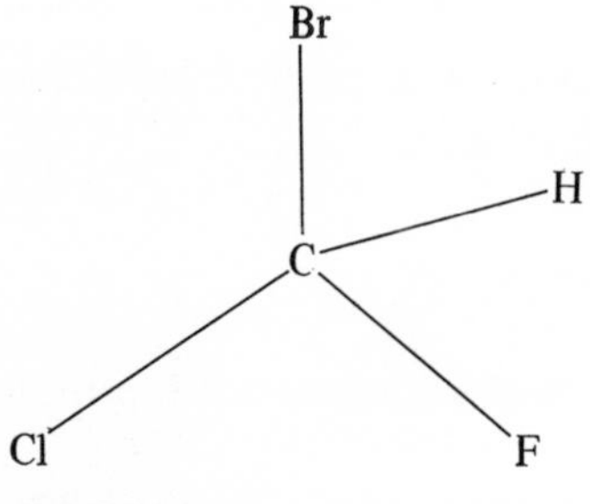

FIGURE 3.1
Example of the C_1 point group

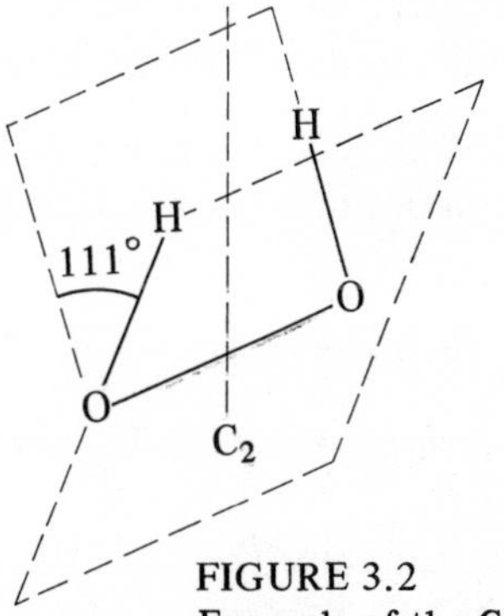

FIGURE 3.2
Example of the C_2 point group

† These elements are not *symmetry* elements but elements required by the definition of a group (see section 3.13).

Examples of molecules belonging to the C_n point groups with $n > 2$ are very rare.

3.2 S_n point groups

An S_n *point group contains the symmetry element* S_n.
An S_n point group also contains the elements generated from S_n by raising it to the powers 2, 3, 4 . . . $(n-1)$. There must not be a plane of symmetry perpendicular to the S_n axis, therefore n must always be even in these point groups.

(*i*) $S_2 \equiv C_i$. The only element of this point group is S_2, which is equivalent to i, a centre of symmetry. An example of a molecule belonging to this point group is an isomer of ClFHC . CHFCl in which the three atoms attached to one carbon are staggered with respect to those attached to the other, and all pairs of identical atoms are *trans* to each other. This molecule is shown in figure 3.3.
C_i is an alternative symbol which is usually used for this point group.

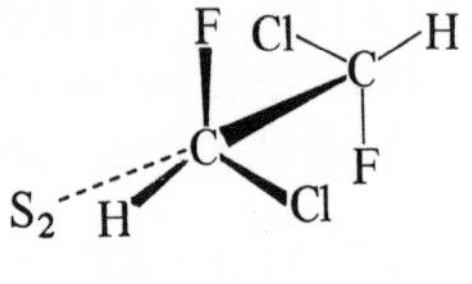

FIGURE 3.3
Example of the C_i point group

(*ii*) S_4. The elements of this point group are S_4, $S_4^2 (= C_2)$, and S_4^3.

Examples of molecules belonging to S_n point groups where $n > 2$ are sufficiently uncommon for them to be of little importance.

3.3 C_{nv} point groups

A C_{nv} *point group contains a* C_n *axis of symmetry and n σ planes of symmetry all containing* C_n.
A C_{nv} point group also contains the elements generated from C_n by raising it to the powers 2, 3, 4 . . . $(n-1)$.

(*i*) $C_{1v} \equiv C_s$. The only element of this point group apart from $C_1 (= I)$ is a plane of symmetry. C_s, as this point group is usually labelled, is a very common group, since any planar molecule, such as phenol (figure 3.4(a)), having no other element of symmetry belongs to it. Some non-planar molecules such as monofluoroallene (figure 3.4(b)) and aniline (figure 2.2(c)) also belong to this point group.

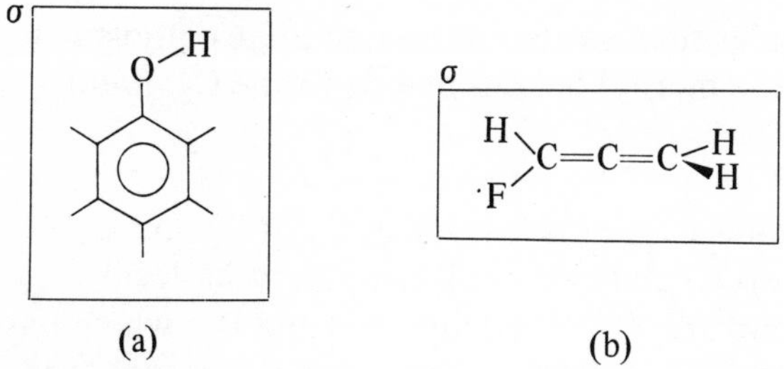

FIGURE 3.4
Examples of the $\mathbf{C_s}$ point group

(*ii*) $\mathbf{C_{2v}}$. This point group contains a C_2 axis and two σ_v planes of symmetry. It is a very common point group, to which water (figure 3.5(a)), *ortho*- and *meta*-dichlorobenzene (figure 3.5(b) (c)), difluoromethane (figure 2.2(a)) and many other molecules belong.

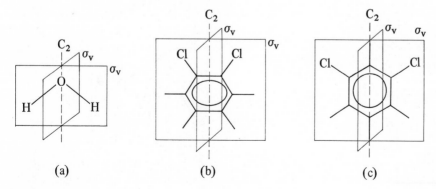

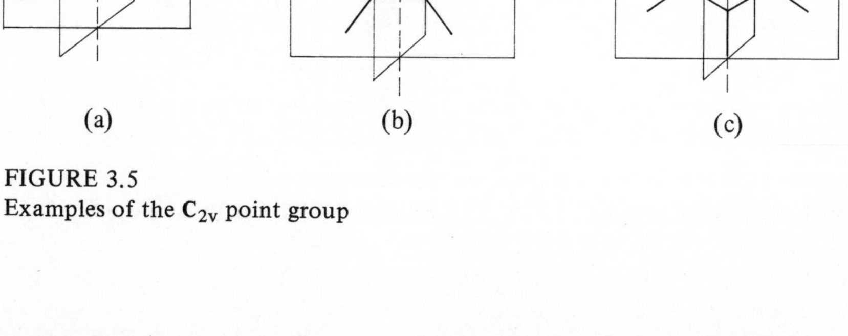

FIGURE 3.5
Examples of the $\mathbf{C_{2v}}$ point group

(*iii*) $\mathbf{C_{3v}}$. As well as the symmetry elements C_3 and three σ_v planes this point group contains the element C_3^2 generated from C_3. Just as the $\mathbf{C_{2v}}$ point group tends to be used as a model point group which contains no C_n axis with $n > 2$, the $\mathbf{C_{3v}}$ point group is often used as a model point group which does contain such an axis.

All trigonal pyramidal molecules such as ammonia (figure 3.6), phosphine (PH_3), arsine (AsH_3), NF_3 and the methyl halides belong to the $\mathbf{C_{3v}}$ point group.

(*iv*) $\mathbf{C_{4v}}$. This point group contains a C_4 axis (as well as $C_4^2 = C_2$ and $C_4^3 = C_4^{-1}$) and four σ planes, two σ_v and two σ_d. Examples of molecules belonging to this point group are $XeOF_4$ (figure 3.7) and IF_5 which has a square pyramidal structure with fluorine atoms at all the corners and the iodine atom at the centre of the base.

Examples of molecules belonging to $\mathbf{C_{nv}}$ point groups where $n > 4$ are very rare, except for the important point group $\mathbf{C_{\infty v}}$.

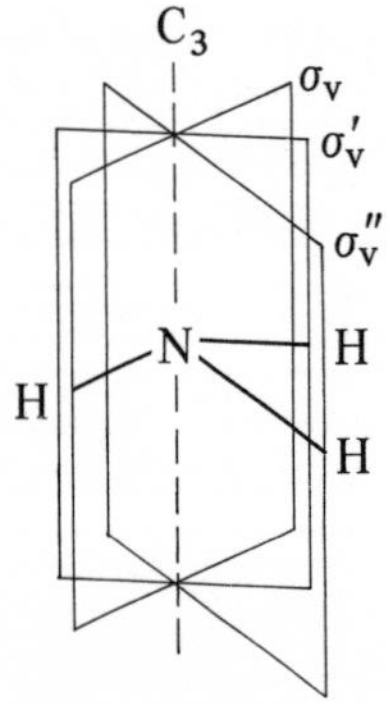

FIGURE 3.6
Example of the $\mathbf{C_{3v}}$ point group

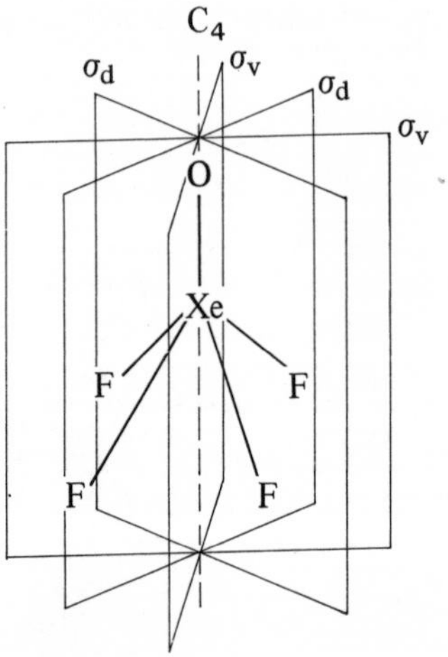

FIGURE 3.7
Example of the $\mathbf{C_{4v}}$ point group

(*v*) $C_{\infty v}$. This point group contains an infinite number of σ_v planes and a C_∞^ϕ axis, where ϕ represents any arbitrary angle of rotation, (as well as $C_\infty^{2\phi}$, $C_\infty^{3\phi}$ . . . and $C_\infty^{-2\phi}$, $C_\infty^{-3\phi}$ . . .). All linear unsymmetrical molecules such as HCN (figure 2.1(e)) belong to this point group. Any plane containing the C_∞ axis is a σ_v plane of symmetry.

3.4 D_n point groups

A $\mathbf{D}_n$ *point group contains the symmetry elements* C_n *and n* C_2 *axes. The* C_2 *axes are perpendicular to* C_n *and at equal angles to each other.*
A $\mathbf{D}_n$ point group also contains the elements generated from C_n by raising it to the powers 2, 3, 4 . . . $(n-1)$.

(*i*) $\mathbf{D}_1 \equiv \mathbf{C}_2$. This point group contains $C_1(= I)$ and one C_2 axis: it is equivalent, therefore, to the $\mathbf{C}_2$ point group.
(*ii*) $\mathbf{D}_2$. This point group contains three mutually perpendicular C_2 axes.

In general a $\mathbf{D}_n$ molecule is obtained by bringing together two identical $\mathbf{C}_{nv}$ molecules or fragments back-to-back, in such a way that one is staggered with respect to the other by any angle other than $m\pi/n$, where m is also any integer. For example each of the two CH_2 fragments of ethylene ($H_2C=CH_2$) belong to the $\mathbf{C}_{2v}$ point group. If the two fragments are brought together to form ethylene in such a way that the angle between the two CH_2 planes is anything other than $m\pi/2$, the molecule belongs to the $\mathbf{D}_2$ point group (figure 3.8). In its ground electronic state ethylene is planar, but in one of its excited states the CH_2 groups are staggered at an angle of less than $\pi/2$ and in this state it belongs to the $\mathbf{D}_2$ point group.

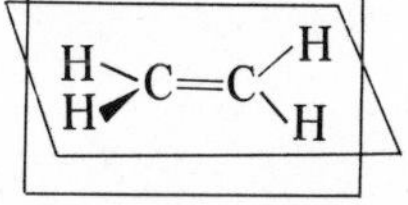

FIGURE 3.8
Example of the $\mathbf{D}_2$ point group

(*iii*) $\mathbf{D}_3$. If two CH_3 groups (each belonging to the C_{3v} point group) are brought together to form ethane ($CH_3.CH_3$) in such a way that the two groups are staggered with respect to each other by any angle other than $m\pi/3$, the molecule belongs to the $\mathbf{D}_3$ point group. However, no state of ethane is known in which it belongs to this point group.

The $\mathbf{D}_3$ point group contains a C_3 axis and three C_2 axes (as well as $C_3^2 = C_3^{-1}$). The C_2 axes are perpendicular to C_3 and at equal angles to each other.

Examples of molecules belonging to $\mathbf{D}_n$ point groups where $n > 3$ are even more difficult to find.

3.5 C_{nh} point groups

A $\mathbf{C}_{nh}$ *point group contains the symmetry element* C_n *and a* σ_h *plane perpendicular to the* C_n *axis. If n is even the point group necessarily contains a centre of symmetry* i.

A $\mathbf{C}_{nh}$ point group also contains the elements generated from C_n by raising it to the powers 2, 3, 4 . . . $(n - 1)$ and also S_n^q elements generated from the C_n^r (where $r = 2, 3, \ldots (n - 1)$) by multiplying them by σ_h. For example, the $\mathbf{C}_{5h}$ point group contains the elements C_5, C_5^2, C_5^3, C_5^4 and also, therefore, contains

$$\sigma_h \times C_5 = S_5$$
$$\sigma_h \times C_5^2 = S_5^7$$
$$\sigma_h \times C_5^3 = S_5^3$$
$$\sigma_h \times C_5^4 = S_5^9$$

The $\mathbf{C}_{6h}$ point group contains the elements C_6, $C_3(= C_6^2)$, $C_2(= C_6^3)$, $C_3^2(= C_6^4)$, C_6^5 and also contains

$$\sigma_h \times C_6 = S_6$$
$$\sigma_h \times C_3 = S_3$$
$$\sigma_h \times C_2 = i$$
$$\sigma_h \times C_3^2 = S_3^5$$
$$\sigma_h \times C_6^5 = S_6^5$$

(*i*) $\mathbf{C}_{1h} \equiv \mathbf{C}_{1v} \equiv \mathbf{C}_s$. A molecule belonging to the $\mathbf{C}_{1h}$ point group has only one plane of symmetry. It is arbitrary whether this is labelled σ_v or σ_h and therefore $\mathbf{C}_{1h} \equiv \mathbf{C}_{1v}$, but the symbol $\mathbf{C}_s$ is usually used.

(*ii*) C_{2h}. This point group has a C_2 axis, a σ_h plane and a centre of symmetry i. Examples of molecules belonging to this point group are glyoxal (figure 3.9), *trans*-difluoroethylene (figure 2.3(c)) and 1,4-difluoro-2,5-dichlorobenzene.

(*iii*) C_{3h}. Orthoboric acid, $B(OH)_3$, shown in figure 3.10, belongs to this rather unusual point group. It has a C_3 axis and a σ_h plane (as well as C_3^2, S_3, and S_3^5).

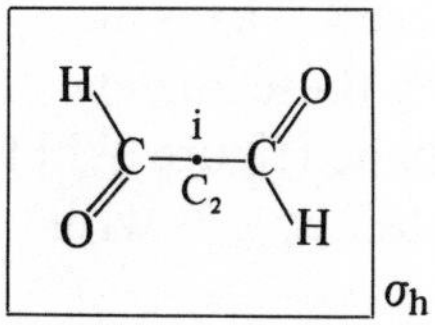

FIGURE 3.9
Example of the C_{2h} point group

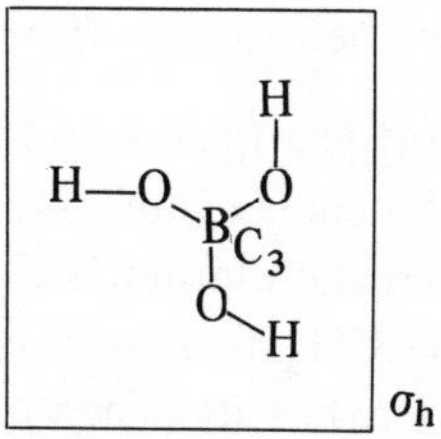

FIGURE 3.10
Example of the C_{3h} point group

Examples of molecules belonging to C_{nh} point groups where $n > 3$ are very rare.

3.6 D_{nd} point groups

A D_{nd} *point group contains the symmetry elements* C_n, S_{2n}, *n* C_2 *axes perpendicular to* C_n *and at equal angles to each other, and n* σ_d *planes bisecting the angles between the* C_2 *axes. If n is odd the point group necessarily contains a centre of symmetry* i.

A D_{nd} point group also contains the elements generated from C_n by raising it to the powers 2, 3, 4 . . . $(n - 1)$, together with the elements generated from S_{2n} by raising it to the powers 3, 5, 7 . . . $(2n - 1)$. When n is odd, S_{2n}^n is equal to i.

A molecule belonging to a D_{nd} point group can be regarded as consisting of two identical fragments of C_{nv} symmetry, staggered at an angle of π/n with respect to each other.

(*i*) $\mathbf{D_{2d}}$. This point group has three C_2 axes, of which two (labelled C_2') are equivalent, one S_4 axis and two σ_d planes bisecting the two C_2' axes. The C_2' axes are perpendicular to each other and to the S_4 axis. (There is also an S_4^3 element.)

Allene (figure 2.2(d)) is an example of a molecule belonging to the $\mathbf{D_{2d}}$ point group. It can be regarded as two fragments of $\mathbf{C_{2v}}$ symmetry oriented at an angle of $\pi/2$ with respect to each other. Ethylene, in one of its excited electronic states, has one CH_2 fragment twisted by $\pi/2$ with respect to the other and therefore belongs, in this state, to the $\mathbf{D_{2d}}$ point group.

(*ii*) $\mathbf{D_{3d}}$. This point group has one C_3 axis, one S_6 axis, three C_2 axes at equal angles to each other and three σ_d planes bisecting these angles. (There are also the elements C_3^2, $S_6^3 = i$, and S_6^5). Ethane (figure 3.11), which has one CH_3 fragment (of $\mathbf{C_{3v}}$ symmetry) oriented by $\pi/3$ with respect to the other, belongs to this point group.

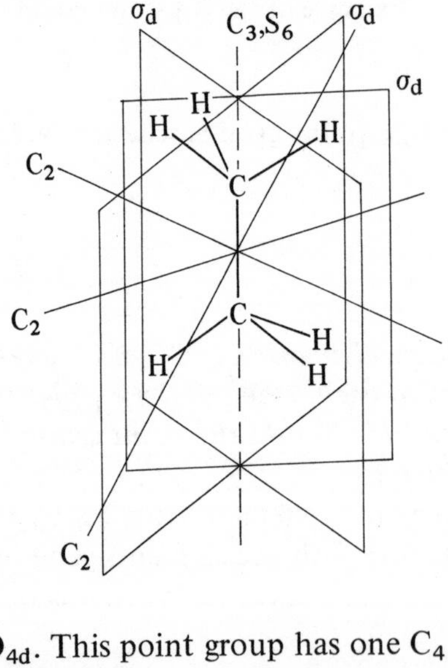

FIGURE 3.11
Example of the $\mathbf{D_{3d}}$ point group

(*iii*) $\mathbf{D_{4d}}$. This point group has one C_4 axis, one S_8 axis, four C_2 axes at equal angles to each other and four σ_d planes bisecting these angles. (There are also the elements $C_4^2 = C_2$, C_4^3, S_8^3, S_8^5 and S_8^7). Any molecule belonging to this point group would have the form of two square pyramids (each of $\mathbf{C}_{4v}$ symmetry), with their apices touching and one oriented by $\pi/4$ relative to the other (figure 3.12).

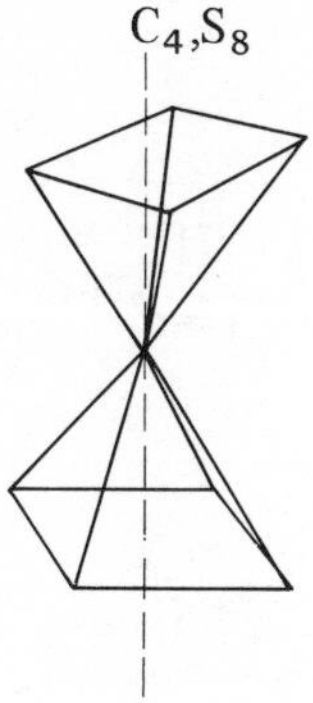

FIGURE 3.12
Example of the $\mathbf{D}_{4d}$ point group

(*iv*) $\mathbf{D}_{5d}$. Ferrocene (figure 2.4(c)) is an example of a molecule belonging to the $\mathbf{D}_{5d}$ point group. This group has one C_5 axis, one S_{10} axis, five C_2 axes at equal angles to each other and five σ_d planes bisecting these angles. (There are also the elements $C_5{}^2$, $C_5{}^3$, $C_5{}^4$, $S_{10}{}^3$, $S_{10}{}^5 = i$, $S_{10}{}^7$, $S_{10}{}^9$). One of the C_5H_5 rings is oriented by $\pi/5$ relative to the other.

Examples of molecules belonging to $\mathbf{D}_{nd}$ point groups with $n > 5$ are rare.

3.7 D_{nh} point groups

A $\mathbf{D}_{nh}$ *point group contains the symmetry elements* C_n*, n* C_2 *axes perpendicular to* C_n *and at equal angles to each other, a* σ_h *plane and n other* σ *planes. If n is even, the point group necessarily contains a centre of symmetry* i.

A $\mathbf{D}_{nh}$ point group also contains the elements generated from C_n by raising it to the powers 2, 3, 4 . . . $(n-1)$ together with, as in the $\mathbf{C}_{nh}$ point groups, all the $S_n{}^q$ elements generated from $\sigma_h \times C_n{}^r$ where $r = 1, 2, \ldots (n-1)$.

A $\mathbf{D}_{nh}$ point group is related to the corresponding $\mathbf{C}_{nv}$ point group by the addition of a σ_h plane of symmetry.

(*i*) $\mathbf{D}_{2h}$. This point group has three mutually perpendicular C_2 axes, three σ planes and a centre of symmetry i. Since all the C_2 axes are equivalent the 'v' or 'h', subscripts for the symbols for the planes of symmetry are not relevant. Ethylene in its planar ground state configuration (figure 3.13(a)) belongs to this point group as do naphthalene (figure 2.3(b)), *p*-difluorobenzene, diborane (figure 3.13(b)) and many other molecules.

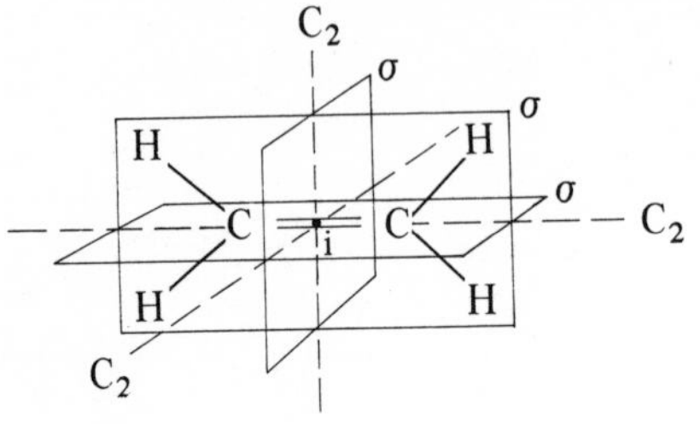

(a)

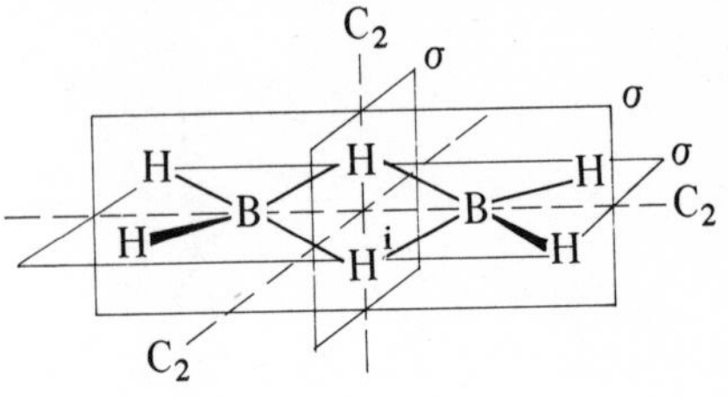

(b)

FIGURE 3.13
Examples of the $\mathbf{D_{2h}}$ point group

(*ii*) $\mathbf{D_{3h}}$. This point group has a C_3 axis, three C_2 axes perpendicular to C_3, three σ_v planes and a σ_h plane. (In addition are the elements C_3^2, S_3 and S_3^2). 1,3,5-Trifluorobenzene (figure 3.14) and boron trifluoride (figure 2.2(b)) belong to this point group.

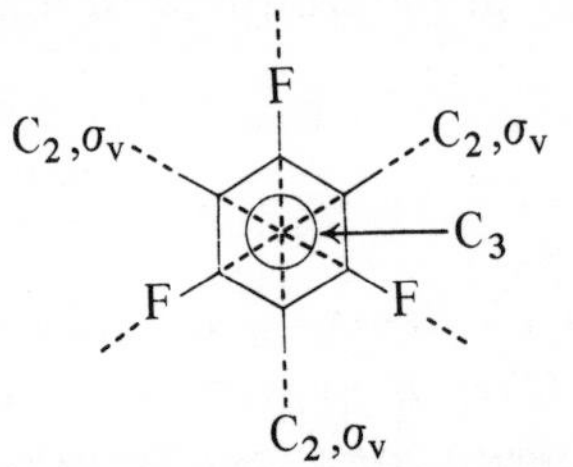

FIGURE 3.14
Example of the $\mathbf{D_{3h}}$ point group

(*iii*) $\mathbf{D_{4h}}$. Any molecule having the symmetry of a square belongs to this point group, which has a C_4 axis, four C_2 axes perpendicular to C_4, four planes of symmetry (two σ_v and two σ_d) and a σ_h plane. (In addition it has the elements $C_4^2 = C_2$, C_4^3, S_4, $S_4^2 = i$, S_4^3). The planar $[PtCl_4]^{2-}$ ion (figure 3.15) is an example of this point group.

(*iv*) $\mathbf{D_{5h}}$. The cyclopentadienyl radical C_5H_5 (cyclopentadiene with one hydrogen atom removed) is a planar molecule (figure 3.16) and belongs to this point group. It has a C_5 axis, five C_2 axes, five σ_v planes and a σ_h plane. (It also has the elements C_5^2, C_5^3, C_5^4, S_5, S_5^2, S_5^3 and S_5^4).

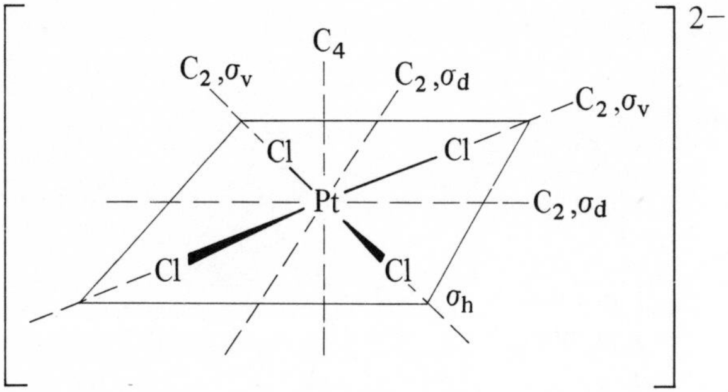

FIGURE 3.15
Example of the $\mathbf{D_{4h}}$ point group

(v) $\mathbf{D_{6h}}$. This point group is of special importance because benzene (figure 3.17), one of the most important aromatic molecules, belongs to it. The point group contains the symmetry elements C_6, three σ_v planes, three σ_d planes, six C_2 axes and a σ_h plane (as well as C_6^2, $C_6^3 = C_2$, C_6^4, C_6^5, S_6, S_6^2, $S_6^3 = i$, S_6^4 and S_6^5).

Molecules belonging to $\mathbf{D}_{nh}$ point groups with $n > 6$ are rare except in the case of the $\mathbf{D_{\infty h}}$ group.

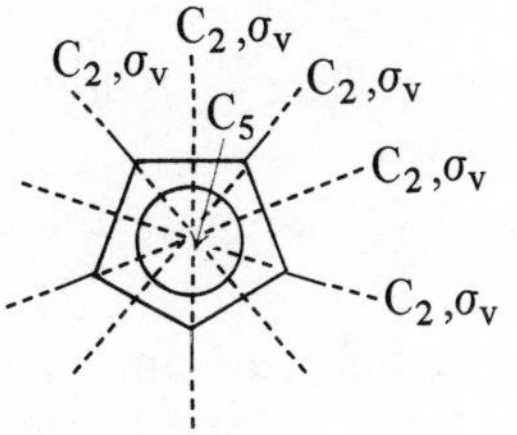

FIGURE 3.16
Example of the $\mathbf{D_{5h}}$ point group

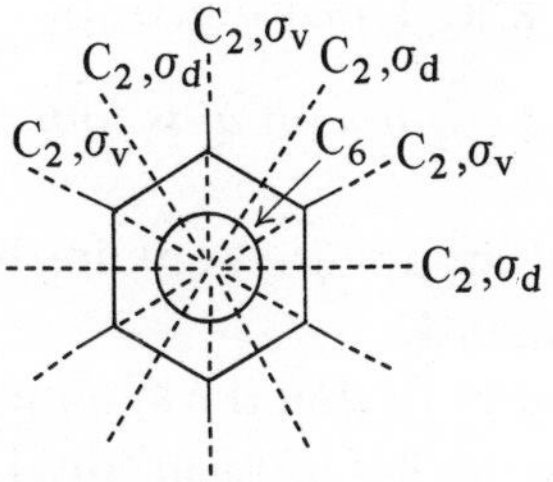

FIGURE 3.17
Example of the $\mathbf{D_{6h}}$ point group

(vi) $\mathbf{D_{\infty h}}$. This point group has a C_∞^ϕ axis, an infinite number of C_2 axes, an infinite number of σ_v planes, a σ_h plane and a centre of symmetry i (as well as $C_\infty^{2\phi}$, $C_\infty^{3\phi}$ . . ., $C_\infty^{-\phi}$, $C_\infty^{-2\phi}$, $C_\infty^{-3\phi}$ . . ., S_∞^ϕ, $S_\infty^{2\phi}$, $S_\infty^{3\phi}$ . . ., and $S_\infty^{-\phi}$, $S_\infty^{-2\phi}$, $S_\infty^{-3\phi}$ . . .). All homonuclear diatomic molecules such as nitrogen, and all linear polyatomic molecules having a centre of symmetry, such as acetylene (figure 3.18) belong to the $\mathbf{D_{\infty h}}$ point group.

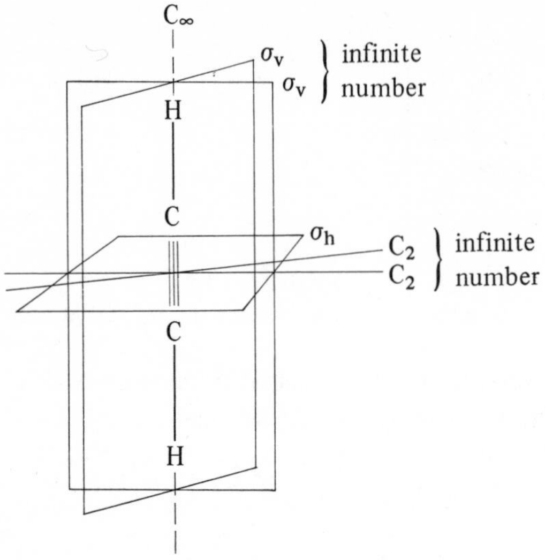

FIGURE 3.18
Example of the $\mathbf{D}_{\infty h}$ point group

3.8 T_d and T point groups

The $\mathbf{T_d}$ *point group contains four* C_3 *axes, three* C_2 *axes and six* σ_d *planes of symmetry.*
The $\mathbf{T_d}$ point group also contains four C_3^2 elements, four S_4 elements and four S_4^3 elements.

All molecules having the symmetry of a tetrahedron such as carbon tetrachloride, nickel tetracarbonyl and methane (figure 3.19) belong to the $\mathbf{T_d}$ point group. The symmetry elements of methane, for example, are most easily shown by considering the cube at four corners of which are, as in figure 3.19, the four hydrogen atoms. The four C_3 axes are the four diagonals of the cube, the three C_2 axes are the three lines joining centre points of opposite faces, and the six σ_d planes are the six planes formed by joining all pairs of diagonally opposite edges of the cube.
The **T** *point group contains four* C_3 *axes and three* C_2 *axes of symmetry.*
The **T** point group also contains four C_3^2 elements.

A molecule belonging to this point group would be essentially tetrahedral in shape, but there would have to be identical groups of atoms at each vertex of the tetrahedron, with each group being oriented so as to destroy the σ_d

planes of the $\mathbf{T_d}$ point group while retaining the C_2 axes. No definite examples of molecules belonging to this point group are known.

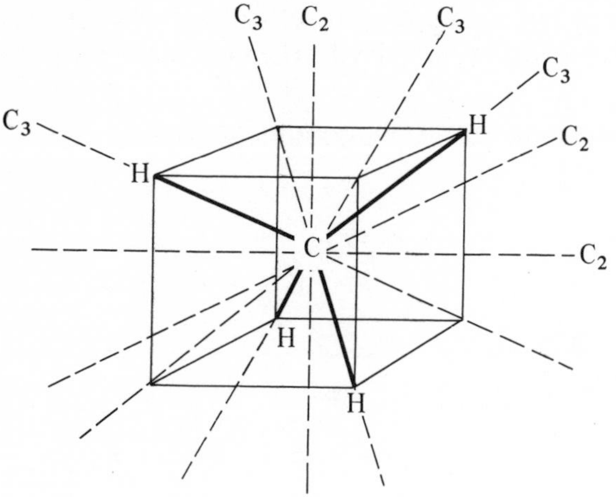

FIGURE 3.19
Example of the $\mathbf{T_d}$ point group

3.9 O_h and O point groups

The $\mathbf{O_h}$ *point group contains three* C_4 *axes, four* C_3 *axes, six* C_2 *axes, three* σ_h *planes, six* σ_d *planes and a centre of symmetry* i.
The $\mathbf{O_h}$ point group also contains three $C_4^2 = C_2$, three C_4^3, three S_4, three S_4^3, four S_6, and four S_6^5 elements.

Any regular octahedral molecule such as the ferricyanide ion $Fe(CN)_6^{3-}$ and SF_6 (figure 3.20) belong to this point group. The main symmetry elements are best demonstrated by considering the octahedron as being inside a cube, such that each vertex of the octahedron is at the centre of a face of the cube. The three C_4 axes join the centre points of opposite faces of the cube, the four C_3 axes are the body diagonals, the six C_2 axes join the centres of pairs of opposite edges, the three σ_h planes are planes halfway between opposite faces and the six σ_d planes join diagonally opposite edges of the cube.

A cube also belongs to this point group.
The **O** *point group contains three* C_4 *axes, four* C_3 *axes and six* C_2 *axes of symmetry.*

The **O** point group also contains three C_4^3, three $C_4^2 (= C_2)$ and four C_3^2 elements.

A molecule belonging to this point group would be essentially octahedral in shape, but there would be groups of atoms at each vertex of the octahedron arranged so that the planes and centre of symmetry of the $\mathbf{O_h}$ point group were lost.

No examples of molecules belonging to this point group are known.

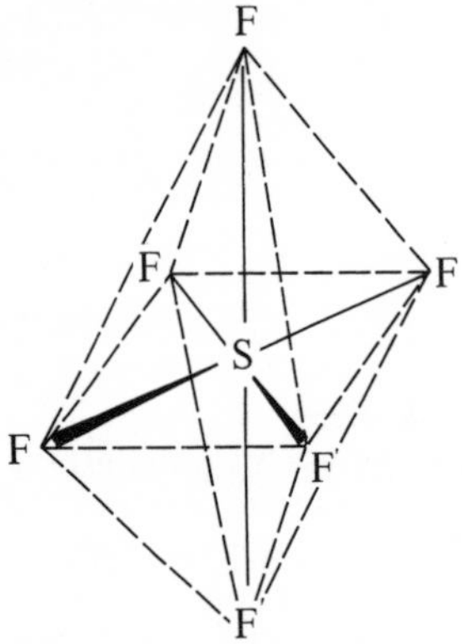

FIGURE 3.20
Example of the $\mathbf{O_h}$ point group

3.10 K_h point group

The $\mathbf{K_h}$ *point group contains an infinite number of* C_∞^ϕ *rotation axes and a centre of symmetry* i.

The $\mathbf{K_h}$ point group also contains the elements generated from all the C_∞^ϕ by raising them to the powers 2, 3, 4 . . . and an infinite number of S_∞^ϕ axes together with the elements generated by raising S_∞^ϕ to the powers 2, 3, 4 . . .

This point group has spherical symmetry and is important because all atoms belong to it.

3.11 Method of assigning a molecule to a point group

Although continual practice in assigning molecules to point groups leads to an almost instinctive ability to do this, it is necessary at first to follow some kind of recipe.

(*i*) If the molecule is linear it can belong only to the $\mathbf{C}_{\infty v}$ or $\mathbf{D}_{\infty h}$ point group. If it has a centre of symmetry it belongs to $\mathbf{D}_{\infty h}$ and to $\mathbf{C}_{\infty v}$ if it has no such centre.

(*ii*) An obvious tetrahedral shape of a molecule implies that it belongs to the $\mathbf{T}_d$ or **T** point group. The main difference between the two is that $\mathbf{T}_d$ contains six σ_d planes of symmetry and **T** does not.

(*iii*) An obvious octahedral shape of a molecule implies that it belongs to the $\mathbf{O}_h$ or **O** point group. These can be distinguished by the presence of a centre of symmetry i in $\mathbf{O}_h$ which is absent in **O**.

(*iv*) If the molecule does not belong to any of the point groups in (*i*)-(*iii*), look for any C_n axes with $n > 1$. If there are no such axes the molecule belongs to the $\mathbf{C}_s$ point group if it has a plane of symmetry σ, to the C_i point group if it has a centre of symmetry i, or to the C_1 point group if it has no element of symmetry at all.

(*v*) If there are one or more C_n axes with $n > 1$, choose the one with the largest value of *n*. In some cases, for example naphthalene ($\mathbf{D}_{2h}$), there are three such axes (C_2). If there is an S_{2n} axis coincident with the C_n axis of highest order (or one of them in the case of three C_2 axes) but there are no other elements of symmetry, except in some cases a centre of symmetry i, the molecule belongs to an $\mathbf{S}_n$ point group.

(*vi*) If there is a C_n axis, a coincident S_{2n} axis and other elements of symmetry, apart from i, then the molecule belongs to a $\mathbf{D}_n$, $\mathbf{D}_{nh}$, $\mathbf{D}_{nd}$, $\mathbf{C}_n$, $\mathbf{C}_{nv}$ or $\mathbf{C}_{nh}$ point group.

(*a*) If there are *n* C_2 axes in a plane perpendicular to the highest order C_n axis the molecule belongs to one of the **D** point groups: if there is a σ_h plane it is $\mathbf{D}_{nh}$, if there are *n* σ_d planes (which are perpendicular to C_n and bisect the angles between C_2 axes) it is $\mathbf{D}_{nd}$ and, if there are no σ_h or σ_d planes, it is $\mathbf{D}_n$.

(*b*) If there are not *n* C_2 axes in a plane perpendicular to the highest order C_n axis the molecule belongs to one of the **C** point groups: if there is a σ_h plane it is $\mathbf{C}_{nh}$, if there are *n* σ_v planes but no σ_h plane it is $\mathbf{C}_{nv}$ and if there are neither σ_h nor σ_v planes then it is $\mathbf{C}_n$.

3.12 Correlation of Schoenflies and Hermann-Maugin symbols for point groups

The Hermann-Maugin system is usually used for the point group symbols for crystal classes, but rarely for the point group symbols of the free molecule. Nevertheless the correlation between Schoenflies and Hermann-Maugin point group symbols is sometimes useful and is given in table 3.1 for the more common point groups.

The Hermann-Maugin system is basedon the fact that only certain of the symmetry elements called the 'generating elements' (see section 3.13) are necessary to define a point group. The Hermann-Maugin symbols comprise these elements. In addition the system uses a solidus separating a symmetry axis symbol from that of a plane, e.g. $4/m$, to imply that the plane is perpendicular to the axis.

TABLE 3.1
Correlation of Schoenflies and Hermann-Maugin symbols for the most common point groups

Schoenflies	*Hermann-Maugin*	*Schoenflies*	*Hermann-Maugin*
$\mathbf{C_1}$	1	$\mathbf{C_{6v}}$	$6mm$
$\mathbf{C_i}$	$\bar{1}$	$\mathbf{D_2}$	222
$\mathbf{C_s}$	m	$\mathbf{D_3}$	32
$\mathbf{C_2}$	2	$\mathbf{D_4}$	422
$\mathbf{C_3}$	3	$\mathbf{D_6}$	622
$\mathbf{C_4}$	4	$\mathbf{D_{2h}}$	mmm
$\mathbf{C_6}$	6	$\mathbf{D_{3h}}$	$\bar{6}m2$
$\mathbf{S_4}$	$\bar{4}$	$\mathbf{D_{4h}}$	$4/mmm$
$\mathbf{C_{2h}}$	$2/m$	$\mathbf{D_{6h}}$	$6/mmm$
$\mathbf{C_{3h}}$	$\bar{6}$	$\mathbf{D_{2d}}$	$\bar{4}\,2m$
$\mathbf{C_{4h}}$	$4/m$	$\mathbf{D_{3d}}$	$\bar{3}\,m$
$\mathbf{C_{6h}}$	$6/m$	$\mathbf{T_d}$	$\bar{4}\,3m$
$\mathbf{C_{2v}}$	$2mm$	$\mathbf{T}$	23
$\mathbf{C_{3v}}$	$3m$	$\mathbf{O_h}$	$m3m$
$\mathbf{C_{4v}}$	$4mm$	$\mathbf{O}$	432

3.13 Properties and definitions in group theory

(i) Properties of a Group

A point group is a rather special example of groups in general. The theory of groups was developed firstly by mathematicians and applied later to the particular problem of molecular symmetry. The elements of all groups obey a rather simple set of rules. With examples taken from point groups, these rules are:

(a) The product of any two elements of a group must also be an element of the group. For example, in the $\mathbf{C}_{2v}$ point group

$$\sigma_v' \times \sigma_v = C_2 \tag{3.1}$$

(b) Products of elements are associative. For example, in the $\mathbf{C}_{2v}$ point group

$$(\sigma_v' \times \sigma_v)C_2 = \sigma_v'(\sigma_v \times C_2) \tag{3.2}$$

where, according to the usual convention, the multiplication inside the brackets is carried out first.

(c) The group contains an element which, when multiplying any other element, leaves it unchanged and which commutes with all the other elements in the group. If two elements commute it means that the result of multiplying the two elements in either order is the same. In general group theory this element is called the unit element, but in point groups it is I, the identity element. For example

$$C_2 \times I = I \times C_2 = C_2 \tag{3.3}$$

(d) Each element of a group has an inverse which is also an element of the group. For example, in the C_{3v} point group, C_3 and C_3^{-1} where

$$C_3^{-1} = C_3^2 \tag{3.4}$$

(ii) Abelian and non-Abelian groups

An Abelian group is one in which all the elements commute with each other i.e. for any two elements P and Q, P x Q = Q x P. This has been discussed in section 2.7 for the elements C_3 and σ_v in the $\mathbf{D}_{3h}$ point group of boron trifluoride. C_3 and σ_v do not commute and the $\mathbf{D}_{3h}$ point group is non-Abelian. The point groups $\mathbf{C}_n$, $\mathbf{S}_n$, $\mathbf{C}_{nh}$, $\mathbf{C}_{2v}$, $\mathbf{D}_2$, $\mathbf{D}_{2h}$ are Abelian. All others are non-Abelian.

(iii) Order of a group

The order of a group is simply the total number of elements in the group: for example, the $\mathbf{C}_{2v}$ point group is of order four ($I, C_2, \sigma_v, \sigma_v'$) and the $\mathbf{C}_{3v}$ point group is of order six ($I, C_3, C_3^2, \sigma_v, \sigma_v', \sigma_v''$).

(iv) Generating elements of a group

If two elements of a group when multiplied together produce a third element of the group, they are said to *generate* the third element. Equation 3.1 shows that, in the $\mathbf{C}_{2v}$ point group, elements σ_v and σ_v' generate C_2. What this really means is that if a molecule has two planes of symmetry perpendicular to each other, it must also have a C_2 axis, a fact which is intuitively obvious. Since I can be generated as follows

$$\sigma_v \times \sigma_v = I \tag{3.5}$$

σ_v and σ_v' can generate all the elements of the group and are called *generating elements* of the $\mathbf{C}_{2v}$ point group. Equally σ_v and C_2 could be taken as generating elements since

$$\begin{aligned} \sigma_v \times C_2 &= \sigma_v' \\ \sigma_v \times \sigma_v &= I \end{aligned} \tag{3.6}$$

In the $\mathbf{C}_{3v}$ point group we can take C_3 and σ_v as generating elements since

$$C_3 \times C_3 = C_3^2 \tag{3.7}$$

and, as illustrated in figure 2.6,

$$\begin{aligned} \sigma_v \times C_3 &= \sigma_v' \\ C_3 \times \sigma_v &= \sigma_v'' \end{aligned} \tag{3.8}$$

and finally, as in equation 3.5,

$$\sigma_v \times \sigma_v = I$$

In many texts on group theory applied to molecular symmetry there is some slight confusion with regard to which elements are listed as defining a particular point group. For example, it is commonly stated that a molecule with a C_3 axis and three σ_v planes belongs to the $\mathbf{C}_{3v}$ point group. Although this sort of definition is a useful one the student must realize that in such a case the elements mentioned are neither the generating elements nor the totality of elements, but some compromise which is necessarily rather arbitrary. Since giving only the generating elements, such as C_3 and σ_v in the $\mathbf{C_{3v}}$ point group, does not usually communicate a picture of the total symmetry of the molecule concerned and the above kind of compromise is arbitrary, the system adopted in this chapter has been to list *all* the elements in the point groups with the most important ones mentioned first in italics.

(v) Classes of elements

Two elements P and Q of a group are said to belong to the same class if there exists a third element R of the group such that

$$P = R^{-1} \times Q \times R \tag{3.9}$$

In addition, elements P and Q are said to be *conjugate* to each other. With the requirement of equation 3.9 it can be shown that the elements of the $\mathbf{C_{3v}}$ point group, namely I, C_3, C_3^2, σ_v, σ_v', and σ_v'', fall into three classes. One class contains all the σ_v planes, another the C_3 and C_3^2 elements, and a third the identity element. Figure 3.21 illustrates the condition

$$C_3 = \sigma_v^{-1} \times C_3^2 \times \sigma_v \tag{3.10}$$

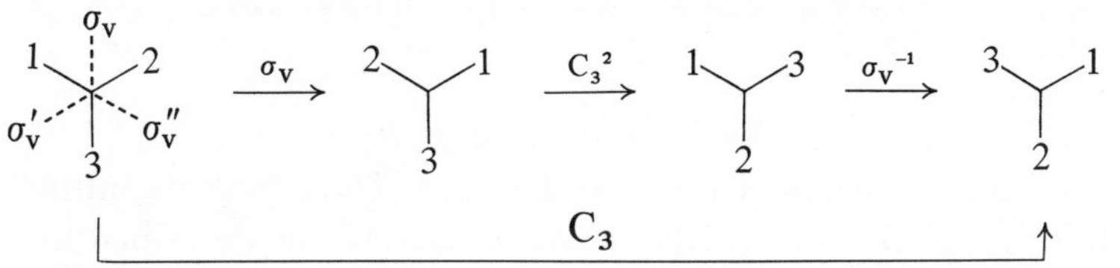

FIGURE 3.21
Illustration of the fact that, in the $\mathbf{C_{3v}}$ point group, $C_3 = \sigma_v^{-1} \times C_3^2 \times \sigma_v$

which shows that C_3 and C_3^2 are in the same class, and figure 3.22 illustrates the conditions

$$\begin{aligned} \sigma_v'' &= C_3^{-1} \times \sigma_v' \times C_3 \\ \sigma_v &= C_3^{-2} \times \sigma_v' \times C_3^2 \end{aligned} \tag{3.11}$$

which shows that σ_v, σ_v', and σ_v'' are in the same class. As further examples, the elements of the $\mathbf{C_{2v}}$ point group, I, C_2, σ_v, σ_v', are all in separate classes and the elements of the $\mathbf{D_{2d}}$ point group, I, S_4, S_4^3, C_2, C_2' (two), σ_d (two) fall into five classes in the following way: I, $2S_4$, C_2, $2C_2'$, $2\sigma_d$.

It is useful to remember that (a) if a point group contains no axis of order higher than two, all the elements are in separate classes, (b) a centre of symmetry i, a σ_h plane, and the identity element are each always in a class by themselves.

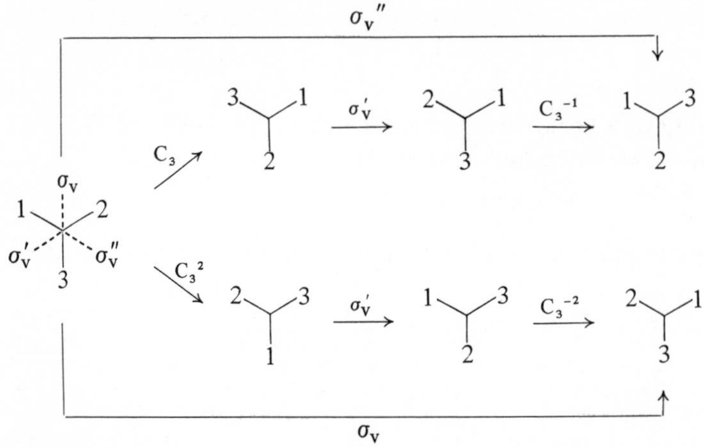

FIGURE 3.22
Illustration of the fact that, in the $\mathbf{C_{3v}}$ point group, $\sigma_v'' = C_3^{-1} \times \sigma_v' \times C_3$ and $\sigma_v = C_3^{-2} \times \sigma_v' \times C_3^2$

(vi) Comparison of a point group with a numerical group

The $\mathbf{C_4}$ point group is an Abelian group of order four. The complete multiplication table for all the elements of the group is obtained as described in section 2.7 and is given in table 3.2. The set of elements 1, – 1, *i*, –*i* is also

TABLE 3.2
Multiplication table for $\mathbf{C_4}$ point group

	I	C_2	C_4	C_4^3
I	I	C_2	C_4	C_4^3
C_2	C_2	I	C_4^3	C_4
C_4	C_4	C_4^3	C_2	I
C_4^3	C_4^3	C_4	I	C_2

an Abelian group of order four. The multiplication table for the elements of the group, given in table 3.3, demonstrates that multiplication of any two elements always produces an element of the group. The elements are clearly associative and commute with each other and 1 is the unit element of this group.

Because of the one-to-one correlation of the elements in the $\mathbf{C}_4$ and 1, -1, i, $-i$ groups, as demonstrated by the multiplication tables 3.2 and 3.3, the groups are said to be *isomorphous.*

TABLE 3.3
Multiplication table for the group 1, -1, i, $-i$.

	1	-1	i	$-i$
1	1	-1	i	$-i$
-1	-1	1	$-i$	i
i	i	$-i$	-1	1
$-i$	$-i$	i	1	-1

Point group character tables 4

So far we have been concerned only with the symmetry properties of the equilibrium configuration of the nuclei of a molecule. However, many molecular properties such as the rotational, vibrational, electronic, nuclear spin or electron spin wave functions, the components of the dipole moment and translations of the molecule, may have lower symmetry than that of the equilibrium nuclear configuration. We require to classify such properties according to their behaviour with respect to the elements in the point group of the molecule.

Classification of properties according to the elements of the point group is considerably easier when the group does not have a higher than two-fold axis–a *non-degenerate* point group e.g. $\mathbf{C}_{2v}$–than when the group does have such an axis–a *degenerate* point group e.g. $\mathbf{C}_{3v}$. For this reason classification of properties in these two types of point groups will be considered separately.

4.1 Non-degenerate point groups

The highest symmetry which a molecular property may have is that of the equilibrium configuration of the nuclei. In molecules belonging to non-degenerate point groups, properties which have lower symmetry are *changed in sign* by carrying out one or more of the operations of the group. For exampl a vibrational wave function ψ_v may be changed in sign by the operation σ so that

$$\psi_v \xrightarrow{\sigma} (-1)\psi_v \tag{4.1}$$

in which case ψ_v is said to be *antisymmetric* with respect to the operation σ. In addition, -1 is called the *character* of the vibrational wave function with respect to σ. If ψ_v is unchanged by the operation σ, then

$$\psi_v \xrightarrow{\sigma} (+1)\psi_v \tag{4.2}$$

and ψ_v is *symmetric* with respect to σ and its character is + 1.

It is characteristic of non-degenerate point groups that characters may be + 1 or − 1 only. As we shall see in section 4.2 this is not so for degenerate point groups.

Any property can now be classified according to its symmetric or antisymmetric behaviour with respect to all elements in the group. A particular combination of + 1 and − 1 characters with respect to the elements in a point group is known as a *symmetry species* of the point group: it is also known more generally as an *irreducible representation.* If all the characters of the symmetry species are + 1 then it is called the *totally symmetric* species: all other species are *non-totally symmetric.* The characters of all the symmetry species with respect to all the elements form the *character table* of the point group.

(*a*) $\mathbf{C_s}$. The character table for the $\mathbf{C_s}$ point group is a very simple one and easy to derive. This point group has only two elements, I and σ. All properties must be symmetric with respect to I, that is the characters of all the symmetry species with respect to I must be + 1. (This is true for all symmetry species in all non-degenerate point groups.) Properties may have the character + 1 or − 1 with respect to σ. There are therefore only two symmetry species in this point group: they are labelled A′ and A″ which are respectively symmetric and antisymmetric with respect to σ. The labels used for symmetry species are conventional and fairly systematic. Primes are always used to indicate symmetric or antisymmetric behaviour with respect to a σ plane perpendicular to the highest-fold C_n axis (in this case C_1).

The character table of the $\mathbf{C_s}$ point group is given in table 4.1. In addition to the information regarding symmetry species, this table, and all the other character tables given, contains in the righthand column an assignment to symmetry species of the components of the polarizability tensor α, the rotations R about and translations T along the principal cartesian axes. Methods for obtaining the symmetry species of rotations and translations will be discussed in section 4.4 and polarizabilities in section 7.5, but it is worth noting here that the z-axis is taken to be the unique one (if there is one): in the $\mathbf{C_s}$ point group it is taken to be perpendicular to the σ plane. As examples of the assignment of a symmetry species to a particular property, figure 4.1 illustrates two vibrational modes in *o*-fluorochlorobenzene ($\mathbf{C_s}$

INDEX TO CHARACTER TABLES

Table no.	Point group	Table no.	Point group
4.1	C_s	4.23	C_{3h}
4.2	C_i	4.24	C_{4h}
4.3	C_1	4.25	C_{5h}
4.4	C_2	4.26	C_{6h}
4.5	C_3	4.27	D_{2d}
4.6	C_4	4.28	D_{3d}
4.7	C_5	4.29	D_{4d}
4.8	C_6	4.30	D_{5d}
4.9	C_7	4.31	D_{6d}
4.10	C_8	4.32	D_{2h}
4.11	C_{2v}	4.33	D_{3h}
4.12	C_{3v}	4.34	D_{4h}
4.13	C_{4v}	4.35	D_{5h}
4.14	C_{5v}	4.36	D_{6h}
4.15	C_{6v}	4.37	$D_{\infty h}$
4.16	$C_{\infty v}$	4.38	S_4
4.17	D_2	4.39	S_6
4.18	D_3	4.40	S_8
4.19	D_4	4.41	T_d
4.20	D_5	4.42	T
4.21	D_6	4.43	O_h
4.22	C_{2h}	4.44	O
		4.45	K_h

TABLE 4.1

C_s	I	σ		
A′	1	1	T_x, T_y, R_z	$\alpha_{x^2}, \alpha_{y^2}, \alpha_{z^2}, \alpha_{xy}$
A″	1	−1	T_z, R_x, R_y	α_{yz}, α_{xz}

TABLE 4.2

C_i	I	i		
A_g	1	1	R_x, R_y, R_z	$\alpha_{x^2}, \alpha_{y^2}, \alpha_{z^2}, \alpha_{xy}, \alpha_{xz}, \alpha_{yz}$
A_u	1	−1	T_x, T_y, T_z	

TABLE 4.3

C_1	I	
A	1	All R, T, α

TABLE 4.4

C_2	I	C_2		
A	1	1	T_z, R_z	$\alpha_{x^2}, \alpha_{y^2}, \alpha_{z^2}, \alpha_{xy}$
B	1	−1	T_x, T_y, R_x, R_y	α_{yz}, α_{xz}

TABLE 4.5

C_3	I	C_3	C_3^2		
A	1	1	1	T_z, R_z	$\alpha_{x^2+y^2}, \alpha_{z^2}$
E	$\left\{\begin{matrix}1 \\ 1\end{matrix}\right.$	$\begin{matrix}\epsilon \\ \epsilon^*\end{matrix}$	$\left.\begin{matrix}\epsilon^* \\ \epsilon\end{matrix}\right\}$	$(T_x, T_y), (R_x, R_y)$	$(\alpha_{x^2-y^2}, \alpha_{xy}), (\alpha_{yz}, \alpha_{xz})$

$\epsilon = \exp(2\pi i/3)$, $\epsilon^* = \exp(-2\pi i/3)$

TABLE 4.6

C_4	I	C_4	C_2	C_4^3		
A	1	1	1	1	T_z, R_z	$\alpha_{x^2+y^2}, \alpha_{z^2}$
B	1	−1	1	−1		$\alpha_{x^2-y^2}, \alpha_{xy}$
E	$\left\{\begin{matrix}1 \\ 1\end{matrix}\right.$	$\begin{matrix}i \\ -i\end{matrix}$	$\begin{matrix}-1 \\ -1\end{matrix}$	$\left.\begin{matrix}-i \\ i\end{matrix}\right\}$	$(T_x, T_y), (R_x, R_y)$	$(\alpha_{yz}, \alpha_{xz})$

TABLE 4.7

$\mathbf{C_5}$	I	C_5	C_5^2	C_5^3	C_5^4		
A	1	1	1	1	1	T_z, R_z	$\alpha_{x^2+y^2}$, α_{z^2}
E_1	1	ϵ	ϵ^2	ϵ^{2*}	ϵ^*	(T_x, T_y), (R_x, R_y)	$(\alpha_{yz}, \alpha_{xz})$
	1	ϵ^*	ϵ^{2*}	ϵ^2	ϵ		
E_2	1	ϵ^2	ϵ^*	ϵ	ϵ^{2*}		$(\alpha_{x^2-y^2}, \alpha_{xy})$
	1	ϵ^{2*}	ϵ	ϵ^*	ϵ^2		

$\epsilon = \exp(2\pi i/5)$, $\epsilon^* = \exp(-2\pi i/5)$

TABLE 4.8

$\mathbf{C_6}$	I	C_6	C_3	C_2	C_3^2	C_6^5		
A	1	1	1	1	1	1	T_z, R_z	$\alpha_{x^2+y^2}$, α_{z^2}
B	1	−1	1	−1	1	−1		
E_1	1	ϵ	$-\epsilon^*$	−1	$-\epsilon$	ϵ^*	(T_x, T_y), (R_x, R_y)	$(\alpha_{xz}, \alpha_{yz})$
	1	ϵ^*	$-\epsilon$	−1	$-\epsilon^*$	ϵ		
E_2	1	$-\epsilon^*$	$-\epsilon$	1	$-\epsilon^*$	$-\epsilon$		$(\alpha_{x^2-y^2}, \alpha_{xy})$
	1	$-\epsilon$	$-\epsilon^*$	1	$-\epsilon$	$-\epsilon^*$		

$\epsilon = \exp(2\pi i/6)$, $\epsilon^* = \exp(-2\pi i/6)$

TABLE 4.9

$\mathbf{C_7}$	I	C_7	C_7^2	C_7^3	C_7^4	C_7^5	C_7^6		
A	1	1	1	1	1	1	1	T_z, R_z	$\alpha_{x^2+y^2}$, α_{z^2}
E_1	1	ϵ	ϵ^2	ϵ^3	ϵ^{3*}	ϵ^{2*}	ϵ^*	(T_x, T_y), (R_x, R_y)	$(\alpha_{xz}, \alpha_{yz})$
	1	ϵ^*	ϵ^{2*}	ϵ^{3*}	ϵ^3	ϵ^2	ϵ		
E_2	1	ϵ^2	ϵ^{3*}	ϵ^*	ϵ	ϵ^3	ϵ^{2*}		$(\alpha_{x^2-y^2}, \alpha_{xy})$
	1	ϵ^{2*}	ϵ^3	ϵ	ϵ^*	ϵ^{3*}	ϵ^2		
E_3	1	ϵ^3	ϵ^*	ϵ^2	ϵ^{2*}	ϵ	ϵ^{3*}		
	1	ϵ^{3*}	ϵ	ϵ^{2*}	ϵ^2	ϵ^*	ϵ^3		

$\epsilon = \exp(2\pi i/7)$, $\epsilon^* = \exp(-2\pi i/7)$

TABLE 4.10

C_8	I	C_8	C_4	C_8^3	C_2	C_8^5	C_4^3	C_8^7		
A	1	1	1	1	1	1	1	1	T_z, R_z	$\alpha_{x^2+y^2}$, α_{z^2}
B	1	−1	1	−1	1	−1	1	−1		
E_1	$\begin{cases}1\\1\end{cases}$	ϵ ϵ^*	i $-i$	$-\epsilon^*$ $-\epsilon$	−1 −1	$-\epsilon$ $-\epsilon^*$	$-i$ i	ϵ^* ϵ	(T_x, T_y), (R_x, R_y)	$(\alpha_{xz}, \alpha_{yz})$
E_2	$\begin{cases}1\\1\end{cases}$	i $-i$	−1 −1	$-i$ i	1 1	i $-i$	−1 −1	$-i$ i		$(\alpha_{x^2-y^2}, \alpha_{xy})$
E_3	$\begin{cases}1\\1\end{cases}$	$-\epsilon^*$ $-\epsilon$	$-i$ i	ϵ ϵ^*	−1 −1	ϵ^* ϵ	i $-i$	$-\epsilon$ $-\epsilon^*$		

$\epsilon = \exp(2\pi i/8)$, $\epsilon^* = \exp(-2\pi i/8)$

TABLE 4.11

C_{2v}	I	C_2	$\sigma_v(xz)$	$\sigma_v'(yz)$		
A_1	1	1	1	1	T_z	α_{x^2}, α_{y^2}, α_{z^2}
A_2	1	1	−1	−1	R_z	α_{xy}
B_1	1	−1	1	−1	T_x, R_y	α_{xz}
B_2	1	−1	−1	1	T_y, R_x	α_{yz}

TABLE 4.12

C_{3v}	I	$2C_3$	$3\sigma_v$		
A_1	1	1	1	T_z	$\alpha_{x^2+y^2}$, α_{z^2}
A_2	1	1	−1	R_z	
E	2	−1	0	(T_x, T_y), (R_x, R_y)	$(\alpha_{x^2-y^2}, \alpha_{xy})$, $(\alpha_{xz}, \alpha_{yz})$

TABLE 4.13

C_{4v}	I	$2C_4$	C_2	$2\sigma_v$	$2\sigma_d$		
A_1	1	1	1	1	1	T_z	$\alpha_{x^2+y^2}$, α_{z^2}
A_2	1	1	1	−1	−1	R_z	
B_1	1	−1	1	1	−1		$\alpha_{x^2-y^2}$
B_2	1	−1	1	−1	1		α_{xy}
E	2	0	−2	0	0	(T_x, T_y), (R_x, R_y)	$(\alpha_{xz}, \alpha_{yz})$

TABLE 4.14

C_{5v}	I	$2C_5$	$2C_5^2$	$5\sigma_v$		
A_1	1	1	1	1	T_z	$\alpha_{x^2+y^2}$, α_{z^2}
A_2	1	1	1	−1	R_z	
E_1	2	2 cos 72°	2 cos 144°	0	(T_x, T_y), (R_x, R_y)	$(\alpha_{xz}, \alpha_{yz})$
E_2	2	2 cos 144°	2 cos 72°	0		$(\alpha_{x^2-y^2}, \alpha_{xy})$

TABLE 4.15

C_{6v}	I	$2C_6$	$2C_3$	C_2	$3\sigma_v$	$3\sigma_d$		
A_1	1	1	1	1	1	1	T_z	$\alpha_{x^2+y^2}$, α_{z^2}
A_2	1	1	1	1	−1	−1	R_z	
B_1	1	−1	1	−1	1	−1		
B_2	1	−1	1	−1	−1	1		
E_1	2	1	−1	−2	0	0	(T_x, T_y), (R_x, R_y)	$(\alpha_{xz}, \alpha_{yz})$
E_2	2	−1	−1	2	0	0		$(\alpha_{x^2-y^2}, \alpha_{xy})$

TABLE 4.16

$C_{\infty v}$	I	$2C_{\infty}^{\phi}$ . . .	$\infty\sigma_v$		
$A_1 \equiv \Sigma^+$	1	1 . . .	1	T_z	$\alpha_{x^2+y^2}, \alpha_{z^2}$
$A_2 \equiv \Sigma^-$	1	1 . . .	−1	R_z	
$E_1 \equiv \Pi$	2	$2\cos\phi$ . . .	0	$(T_x, T_y), (R_x, R_y)$	$(\alpha_{xz}, \alpha_{yz})$
$E_2 \equiv \Delta$	2	$2\cos 2\phi$ . . .	0		$(\alpha_{x^2-y^2}, \alpha_{xy})$
$E_3 \equiv \Phi$	2	$2\cos 3\phi$ . . .	0		
. . .	.	. . . .	.		
. . .	.	. . . .	.		
. . .	.	. . . .	.		

TABLE 4.17

D_2	I	$C_2(z)$	$C_2(y)$	$C_2(x)$		
A	1	1	1	1		$\alpha_{x^2}, \alpha_{y^2}, \alpha_{z^2}$
B_1	1	1	−1	−1	T_z, R_z	α_{xy}
B_2	1	−1	1	−1	T_y, R_y	α_{xz}
B_3	1	−1	−1	1	T_x, R_x	α_{yz}

TABLE 4.18

D_3	I	$2C_3$	$3C_2$		
A_1	1	1	1		$\alpha_{x^2+y^2}, \alpha_{z^2}$
A_2	1	1	−1	T_z, R_z	
E	2	−1	0	$(T_x, T_y), (R_x, R_y)$	$(\alpha_{x^2-y^2}, \alpha_{xy}), (\alpha_{xz}, \alpha_{yz})$

TABLE 4.19

$\mathbf{D_4}$	I	$2C_4$	$C_2(=C_4^2)$	$2C_2'$	$2C_2''$		
A_1	1	1	1	1	1		$\alpha_{x^2+y^2}$, α_{z^2}
A_2	1	1	1	–1	–1	T_z, R_z	
B_1	1	–1	1	1	–1		$\alpha_{x^2-y^2}$
B_2	1	–1	1	–1	1		α_{xy}
E	2	0	–2	0	0	(T_x, T_y), (R_x, R_y)	$(\alpha_{xz}, \alpha_{yz})$

TABLE 4.20

$\mathbf{D_5}$	I	$2C_5$	$2C_5^2$	$5C_2$		
A_1	1	1	1	1		$\alpha_{x^2+y^2}$, α_{z^2}
A_2	1	1	1	–1	T_z, R_z	
E_1	2	2 cos 72°	2 cos 144°	0	(T_x, T_y), (R_x, R_y)	$(\alpha_{xz}, \alpha_{yz})$
E_2	2	2 cos 144°	2 cos 72°	0		$(\alpha_{x^2-y^2}, \alpha_{xy})$

TABLE 4.21

$\mathbf{D_6}$	I	$2C_6$	$2C_3$	C_2	$3C_2'$	$3C_2''$		
A_1	1	1	1	1	1	1		$\alpha_{x^2+y^2}$, α_{z^2}
A_2	1	1	1	1	–1	–1	T_z, R_z	
B_1	1	–1	1	–1	1	–1		
B_2	1	–1	1	–1	–1	1		
E_1	2	1	–1	–2	0	0	(T_x, T_y), (R_x, R_y)	$(\alpha_{xz}, \alpha_{yz})$
E_2	2	–1	–1	2	0	0		$(\alpha_{x^2-y^2}, \alpha_{xy})$

TABLE 4.22

C_{2h}	I	C_2	i	σ_h		
A_g	1	1	1	1	R_z	$\alpha_{x^2}, \alpha_{y^2}, \alpha_{z^2}, \alpha_{xy}$
B_g	1	−1	1	−1	R_x, R_y	α_{xz}, α_{yz}
A_u	1	1	−1	−1	T_z	
B_u	1	−1	−1	1	T_x, T_y	

TABLE 4.23

C_{3h}	I	C_3	C_3^2	σ_h	S_3	S_3^5		
A′	1	1	1	1	1	1	R_z	$\alpha_{x^2+y^2}, \alpha_{z^2}$
A″	1	1	1	−1	−1	−1	T_z	
E′	{1	ϵ	ϵ^*	1	ϵ	ϵ^*}	(T_x, T_y)	$(\alpha_{x^2-y^2}, \alpha_{xy})$
	{1	ϵ^*	ϵ	1	ϵ^*	ϵ}		
E″	{1	ϵ	ϵ^*	−1	$-\epsilon$	$-\epsilon^*$}	(R_x, R_y)	$(\alpha_{xz}, \alpha_{yz})$
	{1	ϵ^*	ϵ	−1	$-\epsilon^*$	$-\epsilon$}		

$\epsilon = \exp(2\pi i/3)$, $\epsilon^* = \exp(-2\pi i/3)$

TABLE 4.24

C_{4h}	I	C_4	C_2	C_4^3	i	S_4^3	σ_h	S_4		
A_g	1	1	1	1	1	1	1	1	R_z	$\alpha_{x^2+y^2}, \alpha_{z^2}$
B_g	1	−1	1	−1	1	−1	1	−1		$\alpha_{x^2-y^2}, \alpha_{xy}$
E_g	{1	i	−1	$-i$	1	i	−1	$-i$}	(R_x, R_y)	$(\alpha_{xz}, \alpha_{yz})$
	{1	$-i$	−1	i	1	$-i$	−1	i}		
A_u	1	1	1	1	−1	−1	−1	−1	T_z	
B_u	1	−1	1	−1	−1	1	−1	1		
E_u	{1	i	−1	$-i$	−1	$-i$	1	i}	(T_x, T_y)	
	{1	$-i$	−1	i	−1	i	1	$-i$}		

TABLE 4.25

C_{5h}	I	C_5	C_5^2	C_5^3	C_5^4	σ_h	S_5	S_5^7	S_5^3	S_5^9		
A'	1	1	1	1	1	1	1	1	1	1	R_z	$\alpha_{x^2+y^2}, \alpha_{z^2}$
E_1'	1	ϵ	ϵ^2	ϵ^{2*}	ϵ^*	1	ϵ	ϵ^2	ϵ^{2*}	ϵ^*	(T_x, T_y)	
	1	ϵ^*	ϵ^{2*}	ϵ^2	ϵ	1	ϵ^*	ϵ^{2*}	ϵ^2	ϵ		
E_2'	1	ϵ^2	ϵ^*	ϵ	ϵ^{2*}	1	ϵ^2	ϵ^*	ϵ	ϵ^{2*}		$(\alpha_{x^2-y^2}, \alpha_{xy})$
	1	ϵ^{2*}	ϵ	ϵ^*	ϵ^2	1	ϵ^{2*}	ϵ	ϵ^*	ϵ^2		
A''	1	1	1	1	1	−1	−1	−1	−1	−1	T_z	
E_1''	1	ϵ	ϵ^2	ϵ^{2*}	ϵ^*	−1	$-\epsilon$	$-\epsilon^2$	$-\epsilon^{2*}$	$-\epsilon^*$	(R_x, R_y)	$(\alpha_{xz}, \alpha_{yz})$
	1	ϵ^*	ϵ^{2*}	ϵ^2	ϵ	−1	$-\epsilon^*$	$-\epsilon^{2*}$	$-\epsilon^2$	$-\epsilon$		
E_2''	1	ϵ^2	ϵ^*	ϵ	ϵ^{2*}	−1	$-\epsilon^2$	$-\epsilon^*$	$-\epsilon$	$-\epsilon^{2*}$		
	1	ϵ^{2*}	ϵ	ϵ^*	ϵ^2	−1	$-\epsilon^{2*}$	$-\epsilon$	$-\epsilon^*$	$-\epsilon^2$		

$\epsilon = \exp(2\pi i/5)$, $\epsilon^* = \exp(-2\pi i/5)$

TABLE 4.26

C_{6h}	I	C_6	C_3	C_2	C_3^2	C_6^5	i	S_3^5	S_6^5	σ_h	S_6	S_3		
A_g	1	1	1	1	1	1	1	1	1	1	1	1	R_z	$\alpha_{x^2+y^2}, \alpha_{z^2}$
B_g	1	−1	1	−1	1	−1	1	−1	1	−1	1	−1		
E_{1g}	1	ϵ	$-\epsilon^*$	−1	$-\epsilon$	ϵ^*	1	ϵ	$-\epsilon^*$	−1	$-\epsilon$	ϵ^*	(R_x, R_y)	$(\alpha_{xz}, \alpha_{yz})$
	1	ϵ^*	$-\epsilon$	−1	$-\epsilon^*$	ϵ	1	ϵ^*	$-\epsilon$	−1	$-\epsilon^*$	ϵ		
E_{2g}	1	$-\epsilon^*$	$-\epsilon$	1	$-\epsilon^*$	$-\epsilon$	1	$-\epsilon^*$	$-\epsilon$	1	$-\epsilon^*$	$-\epsilon$		$(\alpha_{x^2-y^2}, \alpha_{xy})$
	1	$-\epsilon$	$-\epsilon^*$	1	$-\epsilon$	$-\epsilon^*$	1	$-\epsilon$	$-\epsilon^*$	1	$-\epsilon$	$-\epsilon^*$		
A_u	1	1	1	1	1	1	−1	−1	−1	−1	−1	−1	T_z	
B_u	1	−1	1	−1	1	−1	−1	1	−1	1	−1	1		
E_{1u}	1	ϵ	$-\epsilon^*$	−1	$-\epsilon$	ϵ^*	−1	$-\epsilon$	ϵ^*	1	ϵ	$-\epsilon^*$	(T_x, T_y)	
	1	ϵ^*	$-\epsilon$	−1	$-\epsilon^*$	ϵ	−1	$-\epsilon^*$	ϵ	1	ϵ^*	$-\epsilon$		
E_{2u}	1	$-\epsilon^*$	$-\epsilon$	1	$-\epsilon^*$	$-\epsilon$	−1	ϵ^*	ϵ	−1	ϵ^*	ϵ		
	1	$-\epsilon$	$-\epsilon^*$	1	$-\epsilon$	$-\epsilon^*$	−1	ϵ	ϵ^*	−1	ϵ	ϵ^*		

$\epsilon = \exp(2\pi i/6)$, $\epsilon^* = \exp(-2\pi i/6)$

TABLE 4.27

D_{2d}	I	$2S_4$	C_2	$2C_2'$	$2\sigma_d$		
A_1	1	1	1	1	1		$\alpha_{x^2+y^2}$, α_{z^2}
A_2	1	1	1	−1	−1	R_z	
B_1	1	−1	1	1	−1		$\alpha_{x^2-y^2}$
B_2	1	−1	1	−1	1	T_z	α_{xy}
E	2	0	−2	0	0	(T_x, T_y), (R_x, R_y)	$(\alpha_{xz}, \alpha_{yz})$

TABLE 4.28

D_{3d}	I	$2C_3$	$3C_2$	i	$2S_6$	$3\sigma_d$		
A_{1g}	1	1	1	1	1	1		$\alpha_{x^2+y^2}$, α_{z^2}
A_{2g}	1	1	−1	1	1	−1	R_z	
E_g	2	−1	0	2	−1	0	(R_x, R_y)	$(\alpha_{x^2-y^2}, \alpha_{xy})$, $(\alpha_{xz}, \alpha_{yz})$
A_{1u}	1	1	1	−1	−1	−1		
A_{2u}	1	1	−1	−1	−1	1	T_z	
E_u	2	−1	0	−2	1	0	(T_x, T_y)	

TABLE 4.29

D_{4d}	I	$2S_8$	$2C_4$	$2S_8^3$	C_2	$4C_2'$	$4\sigma_d$		
A_1	1	1	1	1	1	1	1		$\alpha_{x^2+y^2}$, α_{z^2}
A_2	1	1	1	1	1	−1	−1	R_z	
B_1	1	−1	1	−1	1	1	−1		
B_2	1	−1	1	−1	1	−1	1	T_z	
E_1	2	$\sqrt{2}$	0	$-\sqrt{2}$	−2	0	0	(T_x, T_y)	
E_2	2	0	−2	0	2	0	0		$(\alpha_{x^2-y^2}, \alpha_{xy})$
E_3	2	$-\sqrt{2}$	0	$\sqrt{2}$	−2	0	0	(R_x, R_y)	$(\alpha_{xz}, \alpha_{yz})$

TABLE 4.30

D_{5d}	I	$2C_5$	$2C_5^2$	$5C_2$	i	$2S_{10}^3$	$2S_{10}$	$5\sigma_d$		
A_{1g}	1	1	1	1	1	1	1	1		$\alpha_{x^2+y^2}, \alpha_{z^2}$
A_{2g}	1	1	1	−1	1	1	1	−1	R_z	
E_{1g}	2	2 cos 72°	2 cos 144°	0	2	2 cos 72°	2 cos 144°	0	(R_x, R_y)	$(\alpha_{xz}, \alpha_{yz})$
E_{2g}	2	2 cos 144°	2 cos 72°	0	2	2 cos 144°	2 cos 72°	0		$(\alpha_{x^2-y^2}, \alpha_{xy})$
A_{1u}	1	1	1	1	−1	−1	−1	−1		
A_{2u}	1	1	1	−1	−1	−1	−1	1	T_z	
E_{1u}	2	2 cos 72°	2 cos 144°	0	−2	−2 cos 72°	−2 cos 144°	0	(T_x, T_y)	
E_{2u}	2	2 cos 144°	2 cos 72°	0	−2	−2 cos 144°	−2 cos 72°	0		

TABLE 4.31

D_{6d}	I	$2S_{12}$	$2C_6$	$2S_4$	$2C_3$	$2S_{12}^5$	C_2	$6C_2'$	$6\sigma_d$		
A_1	1	1	1	1	1	1	1	1	1		$\alpha_{x^2+y^2}, \alpha_{z^2}$
A_2	1	1	1	1	1	1	1	−1	−1	R_z	
B_1	1	−1	1	−1	1	−1	1	1	−1		
B_2	1	−1	1	−1	1	−1	1	−1	1	T_z	
E_1	2	$\sqrt{3}$	1	0	−1	$-\sqrt{3}$	−2	0	0	(T_x, T_y)	
E_2	2	1	−1	−2	−1	1	2	0	0		$(\alpha_{x^2-y^2}, \alpha_{xy})$
E_3	2	0	−2	0	2	0	−2	0	0		
E_4	2	−1	−1	2	−1	−1	2	0	0		
E_5	2	$-\sqrt{3}$	1	0	−1	$\sqrt{3}$	−2	0	0	(R_x, R_y)	$(\alpha_{xz}, \alpha_{yz})$

TABLE 4.32

D_{2h}	I	$C_2(z)$	$C_2(y)$	$C_2(x)$	i	$\sigma(xy)$	$\sigma(xz)$	$\sigma(yz)$		
A_g	1	1	1	1	1	1	1	1		$\alpha_{x^2}, \alpha_{y^2}, \alpha_{z^2}$
B_{1g}	1	1	−1	−1	1	1	−1	−1	R_z	α_{xy}
B_{2g}	1	−1	1	−1	1	−1	1	−1	R_y	α_{xz}
B_{3g}	1	−1	−1	1	1	−1	−1	1	R_x	α_{yz}
A_u	1	1	1	1	−1	−1	−1	−1		
B_{1u}	1	1	−1	−1	−1	−1	1	1	T_z	
B_{2u}	1	−1	1	−1	−1	1	−1	1	T_y	
B_{3u}	1	−1	−1	1	−1	1	1	−1	T_x	

TABLE 4.33

D_{3h}	I	$2C_3$	$3C_2$	σ_h	$2S_3$	$3\sigma_v$		
A'_1	1	1	1	1	1	1		$\alpha_{x^2+y^2}, \alpha_{z^2}$
A'_2	1	1	−1	1	1	−1	R_z	
E'	2	−1	0	2	−1	0	(T_x, T_y)	$(\alpha_{x^2-y^2}, \alpha_{xy})$
A''_1	1	1	1	−1	−1	−1		
A''_2	1	1	−1	−1	−1	1	T_z	
E''	2	−1	0	−2	1	0	(R_x, R_y)	$(\alpha_{xz}, \alpha_{yz})$

TABLE 4.34

D_{4h}	I	$2C_4$	C_2	$2C_2'$	$2C_2''$	i	$2S_4$	σ_h	$2\sigma_v$	$2\sigma_d$		
A_{1g}	1	1	1	1	1	1	1	1	1	1		$\alpha_{x^2+y^2}, \alpha_{z^2}$
A_{2g}	1	1	1	−1	−1	1	1	1	−1	−1	R_z	
B_{1g}	1	−1	1	1	−1	1	−1	1	1	−1		$\alpha_{x^2-y^2}$
B_{2g}	1	−1	1	−1	1	1	−1	1	−1	1		α_{xy}
E_g	2	0	−2	0	0	2	0	−2	0	0	(R_x, R_y)	$(\alpha_{xz}, {}_{yz})$
A_{1u}	1	1	1	1	1	−1	−1	−1	−1	−1		
A_{2u}	1	1	1	−1	−1	−1	−1	−1	1	1	T_z	
B_{1u}	1	−1	1	1	−1	−1	1	−1	−1	1		
B_{2u}	1	−1	1	−1	1	−1	1	−1	1	−1		
E_u	2	0	−2	0	0	−2	0	2	0	0	(T_x, T_y)	

TABLE 4.35

D_{5h}	I	$2C_5$	$2C_5^2$	$5C_2$	σ_h	$2S_5$	$2S_5^3$	$5\sigma_v$		
A_1'	1	1	1	1	1	1	1	1		$\alpha_{x^2+y^2}, \alpha_{z^2}$
A_2'	1	1	1	−1	1	1	1	−1	R_z	
E_1'	2	2 cos 72°	2 cos 144°	0	2	2 cos 72°	2 cos 144°	0	(T_x, T_y)	
E_2'	2	2 cos 144°	2 cos 72°	0	2	2 cos 144°	2 cos 72°	0		$(\alpha_{x^2-y^2}, \alpha_{xy})$
A_1''	1	1	1	1	−1	−1	−1	−1		
A_2''	1	1	1	−1	−1	−1	−1	1	T_z	
E_1''	2	2 cos 72°	2 cos 144°	0	−2	−2 cos 72°	−2 cos 144°	0	(R_x, R_y)	$(\alpha_{xz}, \alpha_{yz})$
E_2''	2	2 cos 144°	2 cos 72°	0	−2	−2 cos 144°	−2 cos 72°	0		

TABLE 4.36

$\mathbf{D_{6h}}$	I	$2C_6$	$2C_3$	C_2	$3C_2'$	$3C_2''$	i	$2S_3$	$2S_6$	σ_h	$3\sigma_d$	$3\sigma_v$		
A_{1g}	1	1	1	1	1	1	1	1	1	1	1	1		$\alpha_{x^2+y^2}, \alpha_{z^2}$
A_{2g}	1	1	1	1	−1	−1	1	1	1	1	−1	−1	R_z	
B_{1g}	1	−1	1	−1	1	−1	1	−1	1	−1	1	−1		
B_{2g}	1	−1	1	−1	−1	1	1	−1	1	−1	−1	1		
E_{1g}	2	1	−1	−2	0	0	2	1	−1	−2	0	0	(R_x, R_y)	$(\alpha_{xz}, \alpha_{yz})$
E_{2g}	2	−1	−1	2	0	0	2	−1	−1	2	0	0		$(\alpha_{x^2-y^2}, \alpha_{xy})$
A_{1u}	1	1	1	1	1	1	−1	−1	−1	−1	−1	−1		
A_{2u}	1	1	1	1	−1	−1	−1	−1	−1	−1	1	1	T_z	
B_{1u}	1	−1	1	−1	1	−1	−1	1	−1	1	−1	1		
B_{2u}	1	−1	1	−1	−1	1	−1	1	−1	1	1	−1		
E_{1u}	2	1	−1	−2	0	0	−2	−1	1	2	0	0	(T_x, T_y)	
E_{2u}	2	−1	−1	2	0	0	−2	1	1	−2	0	0		

TABLE 4.37

$\mathbf{D_{\infty h}}$	I	$2C_\infty^\phi$	$\infty\sigma_v$	i	$2S_\infty^\phi$	∞C_2		
$A_{1g} \equiv \Sigma_g^+$	1	1 ...	1	1	1 ...	1		$\alpha_{x^2+y^2}, \alpha_{z^2}$
$A_{2g} \equiv \Sigma_g^-$	1	1 ...	−1	1	1 ...	−1	R_z	
$E_{1g} \equiv \Pi_g$	2	$2\cos\phi$ ..	0	2	$-2\cos\phi$ ..	0	(R_x, R_y)	$(\alpha_{xz}, \alpha_{yz})$
$E_{2g} \equiv \Delta_g$	2	$2\cos 2\phi$ ..	0	2	$2\cos 2\phi$ ..	0		$(\alpha_{x^2-y^2}, \alpha_{xy})$
$E_{3g} \equiv \Phi_g$	2	$2\cos 3\phi$ ..	0	2	$-2\cos 3\phi$ ..	0		
⋮	⋮	⋮	⋮	⋮				
$A_{2u} \equiv \Sigma_u^+$	1	1 ...	1	−1	−1 ...	−1	T_z	
$A_{1u} \equiv \Sigma_u^-$	1	1 ...	−1	−1	−1 ...	1		
$E_{1u} \equiv \Pi_u$	2	$2\cos\phi$ ..	0	−2	$2\cos\phi$ ..	0	(T_x, T_y)	
$E_{2u} \equiv \Delta_u$	2	$2\cos 2\phi$ ..	0	−2	$-2\cos 2\phi$ ..	0		
$E_{3u} \equiv \Phi_u$	2	$2\cos 3\phi$ ..	0	−2	$2\cos 3\phi$ ..	0		
⋮	⋮	⋮	⋮	⋮	⋮	⋮		

TABLE 4.38

S_4	I	S_4	C_2	S_4^3		
A	1	1	1	1	R_z	$\alpha_{x^2+y^2}, \alpha_{z^2}$
B	1	−1	1	−1	T_z	$\alpha_{x^2-y^2}, \alpha_{xy}$
E	$\begin{Bmatrix} 1 & i & -1 & -i \\ 1 & -i & -1 & i \end{Bmatrix}$				$(T_x, T_y), (R_x, R_y)$	$(\alpha_{xz}, \alpha_{yz})$

TABLE 4.39

S_6	I	C_3	C_3^2	i	S_6^5	S_6		
A_g	1	1	1	1	1	1	R_z	$\alpha_{x^2+y^2}, \alpha_{z^2}$
E_g	$\begin{Bmatrix} 1 & \epsilon & \epsilon^* & 1 & \epsilon & \epsilon^* \\ 1 & \epsilon^* & \epsilon & 1 & \epsilon^* & \epsilon \end{Bmatrix}$						(R_x, R_y)	$(\alpha_{x^2-y^2}, \alpha_{xy}), (\alpha_{xz}, \alpha_{yz})$
A_u	1	1	1	−1	−1	−1	T_z	
E_u	$\begin{Bmatrix} 1 & \epsilon & \epsilon^* & -1 & -\epsilon & -\epsilon^* \\ 1 & \epsilon^* & \epsilon & -1 & -\epsilon^* & -\epsilon \end{Bmatrix}$						(T_x, T_y)	

$\epsilon = \exp(2\pi i/3)$, $\epsilon^* = \exp(-2\pi i/3)$

TABLE 4.40

S_8	I	S_8	C_4	S_8^3	C_2	S_8^5	C_4^3	S_8^7		
A	1	1	1	1	1	1	1	1	R_z	$\alpha_{x^2+y^2}, \alpha_{z^2}$
B	1	−1	1	−1	1	−1	1	−1	T_z	
E_1	$\begin{Bmatrix} 1 & \epsilon & i & -\epsilon^* & -1 & -\epsilon & -i & \epsilon^* \\ 1 & \epsilon^* & -i & -\epsilon & -1 & -\epsilon^* & i & \epsilon \end{Bmatrix}$								$(T_x, T_y), (R_x, R_y)$	
E_2	$\begin{Bmatrix} 1 & i & -1 & -i & 1 & i & -1 & -i \\ 1 & -i & -1 & i & 1 & -i & -1 & i \end{Bmatrix}$									$(\alpha_{x^2-y^2}, \alpha_{xy})$
E_3	$\begin{Bmatrix} 1 & -\epsilon^* & -i & \epsilon & -1 & \epsilon^* & i & -\epsilon \\ 1 & -\epsilon & i & \epsilon^* & -1 & \epsilon & -i & -\epsilon^* \end{Bmatrix}$									$(\alpha_{xz}, \alpha_{yz})$

$\epsilon = \exp(2\pi i/8)$, $\epsilon^* = \exp(-2\pi i/8)$

TABLE 4.41

T_d	I	$8C_3$	$3C_2$	$6S_4$	$6\sigma_d$		
A_1	1	1	1	1	1		$\alpha_{x^2+y^2+z^2}$
A_2	1	1	1	−1	−1		
E	2	−1	2	0	0		$(\alpha_{2z^2-x^2-y^2}, \alpha_{x^2-y^2})$
$T_1 \equiv F_1$	3	0	−1	1	−1	(R_x, R_y, R_z)	
$T_2 \equiv F_2$	3	0	−1	−1	1	(T_x, T_y, T_z)	$(\alpha_{xy}, \alpha_{xz}, \alpha_{yz})$

TABLE 4.42

T	I	$4C_3$	$4C_3^2$	$3C_2$		
A	1	1	1	1		$\alpha_{x^2+y^2+z^2}$
E	$\begin{Bmatrix}1\\1\end{Bmatrix}$	ϵ ϵ^*	ϵ^* ϵ	$\begin{Bmatrix}1\\1\end{Bmatrix}$		$(\alpha_{2z^2-x^2-y^2}, \alpha_{x^2-y^2})$
$T \equiv F$	3	0	0	−1	$(T_x, T_y, T_z), (R_x, R_y, R_z)$	$(\alpha_{xy}, \alpha_{xz}, \alpha_{yz})$

$\epsilon = \exp(2\pi i/3)$, $\epsilon^* = \exp(-2\pi i/3)$

TABLE 4.43

O_h	I	$8C_3$	$6C_2$	$6C_4$	$3C_2'(=3C_4^2)$	i	$6S_4$	$8S_6$	$3\sigma_h$	$6\sigma_d$		
A_{1g}	1	1	1	1	1	1	1	1	1	1		$\alpha_{x^2+y^2+z^2}$
A_{2g}	1	1	−1	−1	1	1	−1	1	1	−1		
E_g	2	−1	0	0	2	2	0	−1	2	0		$(\alpha_{2z^2-x^2-y^2}, \alpha_{x^2-y^2})$
$T_{1g} \equiv F_{1g}$	3	0	−1	1	−1	3	1	0	−1	−1	(R_x, R_y, R_z)	
$T_{2g} \equiv F_{2g}$	3	0	1	−1	−1	3	−1	0	−1	1		$(\alpha_{xz}, \alpha_{yz}, \alpha_{xy})$
A_{1u}	1	1	1	1	1	−1	−1	−1	−1	−1		
A_{2u}	1	1	−1	−1	1	−1	1	−1	−1	1		
E_u	2	−1	0	0	2	−2	0	1	−2	0		
$T_{1u} \equiv F_{1u}$	3	0	−1	1	−1	−3	−1	0	1	1	(T_x, T_y, T_z)	
$T_{2u} \equiv F_{2u}$	3	0	1	−1	−1	−3	1	0	1	−1		

TABLE 4.44

O	I	$8C_3$	$6C_2$	$6C_4$	$3C_2'(=3C_4^2)$		
A_1	1	1	1	1	1		$\alpha_{x^2+y^2+z^2}$
A_2	1	1	–1	–1	1		
E	2	–1	0	0	2		$(\alpha_{2z^2-x^2-y^2}, \alpha_{x^2-y^2})$
$T_1 \equiv F_1$	3	0	–1	1	–1	$(T_x, T_y, T_z), (R_x, R_y, R_z)$	
$T_2 \equiv F_2$	3	0	1	–1	–1		$(\alpha_{xy}, \alpha_{xz}, \alpha_{yz})$

TABLE 4.45

K_h	I	∞C_∞^ϕ	∞S_∞^ϕ	i		
S_g	1	1	1	1		$\alpha_{x^2} + \alpha_{y^2} + \alpha_{z^2}$
S_u	1	1	−1	−1		
P_g	3	$1 + 2\cos\phi$	$1 - 2\cos\phi$	1	(R_x, R_y, R_z)	
P_u	3	$1 + 2\cos\phi$	$-1 + 2\cos\phi$	−1	(T_x, T_y, T_z)	
D_g	5	$1 + 2\cos\phi + 2\cos 2\phi$	$1 - 2\cos\phi + 2\cos 2\phi$	1		$(\alpha_{2z^2-x^2-y^2}, \alpha_{x^2-y^2}), (\alpha_{xy}, \alpha_{xz}, \alpha_{yz})$
D_u	5	$1 + 2\cos\phi + 2\cos 2\phi$	$-1 + 2\cos\phi - 2\cos 2\phi$	−1		
F_g	7	$1 + 2\cos\phi + 2\cos 2\phi + 2\cos 3\phi$	$1 - 2\cos\phi + 2\cos 2\phi - 2\cos 3\phi$	1		
F_u	7	$1 + 2\cos\phi + 2\cos 2\phi + 2\cos 3\phi$	$-1 + 2\cos\phi - 2\cos 2\phi + 2\cos 3\phi$	−1		
.	.	.	.	.		
.	.	.	.	.		
.	.	.	.	.		
.	.	.	.	.		
.	.	.	.	.		

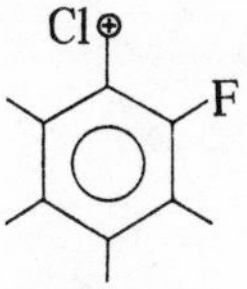

$\Gamma(\Psi_v) = A''$

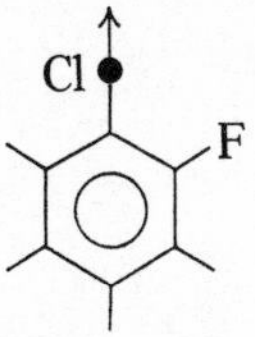

$\Gamma(\Psi_v) = A'$

FIGURE 4.1
Symmetry classification of two vibrational modes of *o*-fluorochlorobenzene

point group) which have a′ and a″ symmetry species.* Another symbolism is introduced here in which Γ stands for 'the symmetry species of . . . ': thus $\Gamma(\psi_v)$ stands for the symmetry species of the vibrational wave function.

(*b*) $\mathbf{C_i}$. This point group resembles the $\mathbf{C_s}$ point group in that it has only two symmetry species: in this case they arise from the only element which is i (apart from I). The symmetry species are labelled A_g and A_u which are respectively symmetric and antisymmetric with respect to the i operation. The g and u labels derive from the German 'gerade' (even) and 'ungerade' (odd) and the labels are always used for symmetry species in point groups having a centre of symmetry i.

The character table is given in table 4.2.

(*c*) $\mathbf{C_1}$. This is a trivial point group having only the I element and only one symmetry species labelled A.

The character table is given in table 4.3.

(*d*) $\mathbf{C_2}$. Having only the element C_2, apart from I, this point group has only two symmetry species. They are labelled A and B and are respectively symmetric and antisymmetric to C_2.

A and B labels for symmetry species are usually used to indicate respectively symmetric or antisymmetric behaviour with respect to rotation about the 'main' axis of the molecule. Some problems arise in the $\mathbf{D_2}$ and $\mathbf{D_{2h}}$ point groups, in each of which there are three mutually perpendicular C_2 axes. In these cases an A symmetry species indicates symmetric behaviour with respect to *all* C_2 operations and B antisymmetric behaviour with respect to

* See section (*e*) for explanation of the use of upper and lower case symbols for symmetry species.

any of the C_2 operations. It is quite useful in trying to remember symmetry species to know why the labels have been chosen in the way they have.

The $\mathbf{C_2}$ character table, in which the C_2 axis is taken to be the z-axis, is given in table 4.4.

(*e*) $\mathbf{C_{2v}}$. The $\mathbf{C_{2v}}$ point group has elements I, C_2, σ_v, σ_v', but just as the elements I and σ_v' can be generated from C_2 and σ_v (section 3.13 (*iv*)) so can the characters of all symmetry species be generated from the characters with respect to the operations C_2 and σ_v. For example, if a symmetry species is symmetric (character + 1) to C_2 and antisymmetric (character − 1) to σ_v, then it must be antisymmetric to σ_v' since $C_2 \times \sigma_v = \sigma_v'$ and $+1 \times -1 = -1$. It can now be seen that the number of symmetry species in the $\mathbf{C_{2v}}$ point group is determined by the number of combinations of + 1 and − 1 characters with respect to the two generating elements, namely four (+ 1, + 1; + 1, − 1; − 1, + 1; − 1, − 1). The symmetry species are labelled A_1, A_2, B_1 and B_2 respectively, the subscripts 1 and 2 indicating (in this point group) respectively symmetry or antisymmetry to the $\sigma_v(xz)$ operation.

In non-degenerate point groups it is a general rule that the number of symmetry species is given by the number of possible combinations of + 1 and − 1 characters among the generating elements. A generally more useful rule is that the number of symmetry species in a point group is the same as the number of classes of elements. This rule applies to *all* point groups.

A new problem arises in the $\mathbf{C_{2v}}$ point group. It is one which has caused much confusion in the past and which to some extent still continues to do so. The B_1 and B_2 symmetry species are distinguished by their symmetry or antisymmetry to reflection across the $\sigma_v(xz)$ plane and the significance of the subscripts 1 and 2 depends entirely on the labelling of the x-, y- and z-axes. Recommendations on choice of axes were made in a report by R. S. Mulliken† which was adopted by the International Union of Pure and Applied Physics and the International Astronomical Union in July, 1954. Whereas in the point groups we have dealt with so far it is sufficient to state that the z-axis is the unique axis, if there is one, in several point groups the x- and y-axes must also be defined conventionally. The recommendation for the $\mathbf{C_{2v}}$ point group is that, according to the general convention, the C_2 axis is

† Mulliken, R. S., 'Report on Notation for the Spectra of Polyatomic Molecules', *J. Chem. Phys.*, 23, 1997 (1955).

labelled the z-axis and that in a planar molecule the axis perpendicular to the molecular plane is labelled the x-axis, as illustrated for formaldehyde in figure 4.2. The importance of axis notation may be realized when it is seen that, for example, an out-of-plane vibration of formaldehyde which has a symmetry species b_1 according to the Mulliken convention† would have a species b_2 if the labelling of the x- and y-axes were interchanged. This type of confusion exists throughout the literature and is made worse by the choice of axes often not being stated. *Mulliken recommends further that authors should always state the choice of axes used even if it is the recommended one.*

In the case of a non-planar C_{2v} molecule such as CF_2Cl_2 the choice of x- and y-axes must be arbitrary and here it is clearly essential to state the choice made.

The C_{2v} character table is given in table 4.11.

(*f*) $\mathbf{D_2}$. Of the four elements, I and three mutually perpendicular C_2 axes, any two of the C_2 axes can be taken as generating elements. Two generating elements give rise to four symmetry species which are labelled A, B_1, B_2, and B_3.

Again the problem of choice of axes arises in principle. In practice molecules in this point group are so rare that the problem does not often have to be resolved. In the case mentioned in section 3.4(*ii*) of twisted ethylene in which

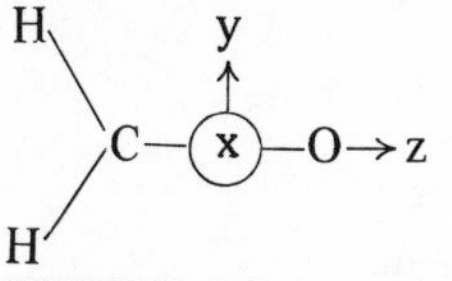

FIGURE 4.2
Axis convention in formaldehyde

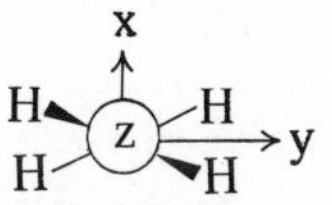

FIGURE 4.3
Axis convention in slightly twisted ethylene

the angle of twist is less than 90° the choice of the C = C axis as the z-axis is an obvious one since this is also the z-axis in planar ethylene which belongs to the D_{2h} point group (see section 4.1 (*h*)). Since, in the D_{2h} point group, the x-axis is perpendicular to the plane of a planar molecule it follows that the axis choice for twisted ethylene should be taken as that shown in figure 4.3.

† Mulliken recommends also the use of *lower case* letters for the symmetry species of a vibrational mode and electronic orbitals. *Upper case* letters are recommended for the symmetry species of wave functions.

The $\mathbf{D}_2$ character table is given in table 4.17. It can be seen from the character table that the B_1, B_2, B_3 species are symmetric to $C_2(z)$, $C_2(y)$, $C_2(x)$ respectively.

(*g*) $\mathbf{C}_{2h}$. This point group has four elements I, C_2, i, and σ_h. It has two generating elements, say C_2 and i, and therefore four symmetry species A_g, B_g, A_u, B_u whose labels follow the rules stated in sections 4.1(*b*) and (*d*).

The $\mathbf{C}_{2h}$ character table, in which the C_2 axis is the z-axis, is given in table 4.22.

(*h*) $\mathbf{D}_{2h}$. The $\mathbf{D}_{2h}$ point group has elements I, $C_2(z)$, $C_2(y)$, $C_2(x)$, i, $\sigma(xy)$, $\sigma(xz)$, and $\sigma(yz)$ of which one choice of generating elements is $C_2(z)$, $C_2(y)$, and i. There are eight possible combinations of characters among the three generating elements and there are therefore eight symmetry species: these are A_g, B_{1g}, B_{2g}, B_{3g}, A_u, B_{1u}, B_{2u}, and B_{3u}. The A, B, u and g have the significance discussed in sections 4.1(*b*) and (*d*) and the subscripts 1, 2, and 3 have the same significance as for the symmetry species in the $\mathbf{D}_2$ point group.

The recommendation of Mulliken is that for planar $\mathbf{D}_{2h}$ molecules the x-axis is perpendicular to the molecular plane and the z-axis passes through the greatest number of atoms or, if this rule is not decisive, cuts the greatest number of bonds. Figure 4.4 illustrates the resulting axis notation in

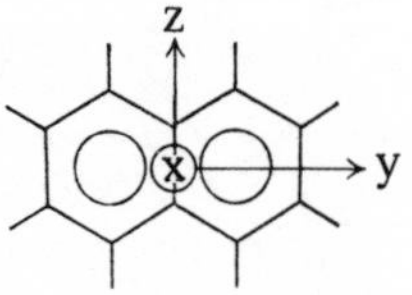

FIGURE 4.4
Axis convention in naphthalene

naphthalene: the z-axis passes through two atoms and the y-axis does not pass through any.

The $\mathbf{D}_{2h}$ character table is given in table 4.32.

4.2 Degenerate point groups

In a molecule belonging to a degenerate point group, that is a point group in which there is a higher than two-fold axis, complications arise due to the fact that some symmetry species may not behave simply in a symmetric or antisymmetric way under an operation. For example, the symmetry species which is labelled E in the $\mathbf{C_{3v}}$ character table (table 4.12) has a character of + 2 under the operation I and a character of 0 under the operation σ_v and, in the $\mathbf{D_{5h}}$ character table (table 4,35), the symmetry species E_1' has a character of 2 cos 72° under the operation C_5.

(*a*) $\mathbf{C_{3v}}$. As shown in section 3.13(*v*) the elements I, C_3, $C_3{}^2$, σ_v, σ_v', σ_v'' of the $\mathbf{C_{3v}}$ point group fall into three classes as given by I, $2C_3$ and $3\sigma_v$. The grouping into classes is important because the character of any symmetry species under all operations in the same class must always be the same: for this reason all the elements in a class are considered together in the $\mathbf{C_{3v}}$ character table given in table 4.12.

A degenerate point group, like a non-degenerate one, has the same number of symmetry species as classes of elements; therefore the $\mathbf{C_{3v}}$ point group has three symmetry species and these are labelled A_1, A_2, and E. As usual A indicates symmetric behaviour to rotation about the main axis of the molecule and the subscripts 1 and 2 indicate respectively symmetric and antisymmetric behaviour with respect to σ_v. E represents a *doubly degenerate* symmetry species. As an example of a molecular property which is doubly degenerate, two vibrations are said to be degenerate if they have the same energy but are described by different wave functions. Triple and higher degeneracies of symmetry species are possible, but fairly unusual. We shall encounter them in only a few point groups whereas doubly degenerate symmetry species are found in all degenerate point groups.

The significance of the E species and in particular how we derive the characters of E under the various operations of the group are best illustrated by examples. The examples used here are the normal vibrations of ammonia which belongs to the $\mathbf{C_{3v}}$ point group. Ammonia has six (= $3N - 6$) normal vibrations which are illustrated in figure 4.5. It is easy to see that the vibrations labelled ν_1 and ν_2 are both a_1 vibrations since they are symmetric with respect to I, C_3, and σ_v. There are no a_2 vibrations in ammonia, but figure 4.6 illustrates an electronic wave function of symmetry species A_2 (Note use of upper case letters for species of wave functions). The vibrations (figure 4.5) labelled ν_{3a} and ν_{3b} are degenerate. They are energetically equivalent, although it is not at all obvious, but it is clear from the form of the vibrations that their wave functions are not identical.

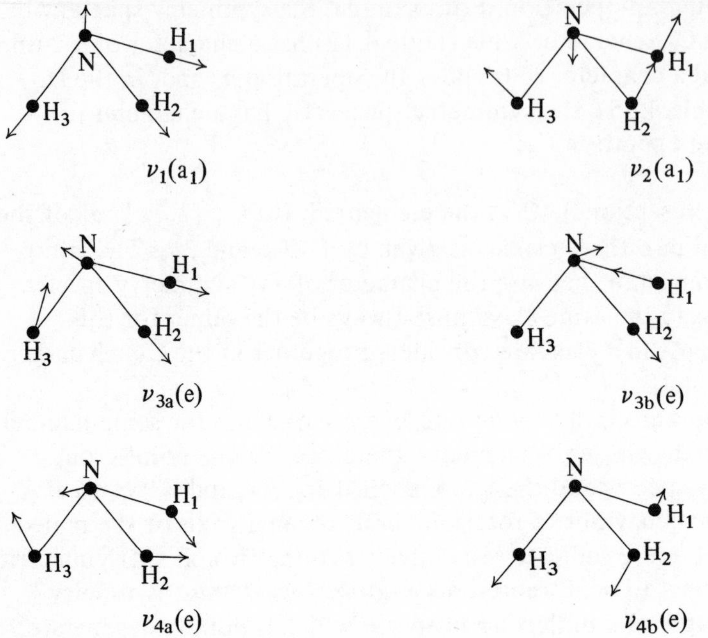

FIGURE 4.5
The normal vibrations of ammonia

The symmetry properties of a normal vibration are the same as those of the corresponding normal co-ordinate, Q. Thus we can write, for example,

$$\begin{aligned} Q_1 &\xrightarrow{C_3} Q_1' = (+1)Q_1 \\ Q_2 &\xrightarrow{\sigma_v} Q_2' = (+1)Q_2 \end{aligned} \tag{4.3}$$

which implies that the normal co-ordinates corresponding to the normal vibrations ν_1 and ν_2 are unchanged by a C_3 or a σ_v operation respectively. However, the normal co-ordinates of a degenerate vibration do not always

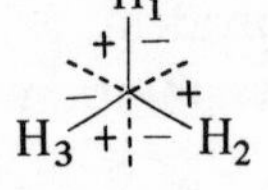

FIGURE 4.6
An A_2 electronic wave function of ammonia.
The dotted lines represent nodes

merely remain unchanged, or change sign, as a result of a symmetry operation: in general they are transformed into a *linear combination* of the normal co-ordinates. In the case of the doubly degenerate vibration ν_3 we can write

$$\begin{aligned} Q_{3a} &\xrightarrow{S} Q'_{3a} = d_{aa}Q_{3a} + d_{ab}Q_{3b} \\ Q_{3b} &\xrightarrow{S} Q'_{3b} = d_{ba}Q_{3a} + d_{bb}Q_{3b} \end{aligned} \tag{4.4}$$

where S represents any symmetry operation and the d_{aa}, etc. are coefficients. If the coefficients are arranged to form a 2 by 2 matrix

$$\begin{pmatrix} d_{aa} & d_{ab} \\ d_{ba} & d_{bb} \end{pmatrix} \tag{4.5}$$

then the quantity $d_{aa} + d_{bb}$ is known as the *trace* of the matrix. It is the trace of the matrix which is the character of the symmetry species of the vibration or any other property concerned. In the case of a non-degenerate symmetry species the matrix is now only 1 by 1 and the trace is simply the only coefficient, for example + 1 in equation 4.3. In the general case of an *n*-fold degenerate symmetry species the character under a symmetry operation will be the trace of an *n* by *n* matrix if the operation transforms the property into a linear combination.

The character of the E species under the I operation is easy to obtain: for example

$$\begin{aligned} Q_{3a} &\xrightarrow{I} Q'_{3a} = 1 . Q_{3a} + 0 . Q_{3b} \\ Q_{3b} &\xrightarrow{I} Q'_{3b} = 0 . Q_{3a} + 1 . Q_{3b} \end{aligned} \tag{4.6}$$

The matrix of coefficients is

$$\begin{pmatrix} 1 & 0 \\ 0 & 1 \end{pmatrix} \tag{4.7}$$

and the trace, which is the character of the E species under the operation I, is 2. In general the character of an *n*-fold degenerate symmetry species under the operation I is *n*.

In molecules belonging to degenerate point groups with σ_v planes of symmetry, it turns out that, of the two components of a doubly degenerate vibration (or other property), it can always be arranged that one is symmetric to reflection across one of the σ_v planes and the other is antisymmetric. For example it can be seen from figure 4.5 that

$$\begin{aligned} Q_{3a} &\xrightarrow{\sigma_v} Q'_{3a} = 1 . Q_{3a} + 0 . Q_{3b} \\ Q_{3b} &\xrightarrow{\sigma_v} Q'_{3b} = 0 . Q_{3a} - 1 . Q_{3b} \end{aligned} \tag{4.8}$$

where σ_v is the plane between atoms H_1 and H_2. The matrix of coefficients is

$$\begin{pmatrix} 1 & 0 \\ 0 & -1 \end{pmatrix} \tag{4.9}$$

and the trace is 0, which is the character of E under the σ_v operation.

The character of the E species under the symmetry operation C_3 is not as easy to obtain. In fact the transformations of the degenerate normal co-ordinates Q_{3a} and Q_{3b} under the C_3 operation are given by

$$\begin{aligned} Q_{3a} &\xrightarrow{C_3} Q'_{3a} = Q_{3a} \cos 2\pi/3 + Q_{3b} \sin 2\pi/3 \\ Q_{3b} &\xrightarrow{C_3} Q'_{3b} = -Q_{3a} \sin 2\pi/3 + Q_{3b} \cos 2\pi/3 \end{aligned} \tag{4.10}$$

The matrix of the coefficients is

$$\begin{pmatrix} \cos 2\pi/3 & \sin 2\pi/3 \\ -\sin 2\pi/3 & \cos 2\pi/3 \end{pmatrix} \text{ or } \begin{pmatrix} -\frac{1}{2} & \frac{\sqrt{3}}{2} \\ -\frac{\sqrt{3}}{2} & -\frac{1}{2} \end{pmatrix} \tag{4.11}$$

and the trace is -1 which is the character of the E species under the C_3 operation.

In general the transformation of doubly degenerate normal co-ordinates Q_{ja} and Q_{jb} under a C_n operation, when $n > 2$, are given by

$$\begin{aligned} Q_{ja} &\xrightarrow{C_n} Q'_{ja} = Q_{ja} \cos 2\pi l/n + Q_{jb} \sin 2\pi l/n \\ Q_{jb} &\xrightarrow{C_n} Q'_{jb} = -Q_{ja} \sin 2\pi l/n + Q_{jb} \cos 2\pi l/n \end{aligned} \tag{4.12}$$

where l is an integer for which $0 < l < n$ and the character of an E species under a C_n operation is given by $2 \cos 2\pi l/n$. The value of l is denoted by a subscript on the E species e.g. for an E_2 species $l = 2$.

An example of a doubly degenerate electronic wave function ψ_e of ammonia with two components ψ_e^a and ψ_e^b is illustrated in figure 4.7. The

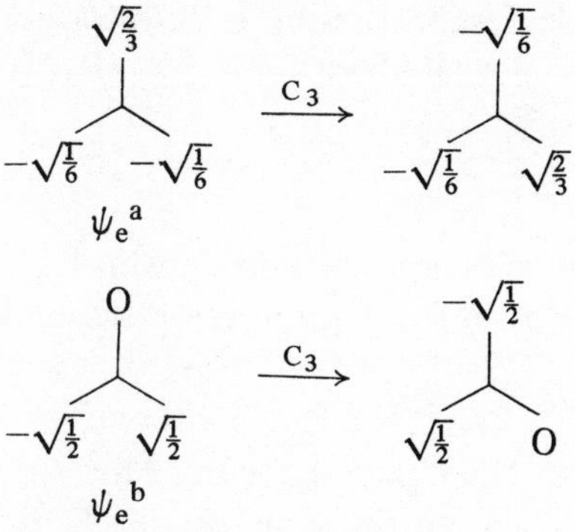

FIGURE 4.7
Effect of C_3 operation on a doubly degenerate electronic wave function of ammonia

numbers attached to the hydrogen atoms represent weightings of the electronic wave function. It can be seen that

$$\begin{aligned} \psi_e^a &\xrightarrow{C_3} (\psi_e^a)' = -\frac{1}{2}\psi_e^a + \frac{\sqrt{3}}{2}\psi_e^b \\ \psi_e^b &\xrightarrow{C_3} (\psi_e^b)' = -\frac{\sqrt{3}}{2}\psi_e^a - \frac{1}{2}\psi_e^b \end{aligned} \tag{4.13}$$

and the matrix representing the transformation is again

$$\begin{pmatrix} -\frac{1}{2} & \frac{\sqrt{3}}{2} \\ -\frac{\sqrt{3}}{2} & -\frac{1}{2} \end{pmatrix}$$

of which the trace is -1.

The axis notation in degenerate point groups is much less of a problem than in non-degenerate point groups. The highest-fold axis, for example C_3 in the $\mathbf{C}_{3v}$ point group, is labelled the z-axis and the other two, which are not distinguishable, are x and y. In the cubic point groups **T**, $\mathbf{T}_d$, **O**, $\mathbf{O}_h$ and the group $\mathbf{K}_h$, none of the three cartesian axes is distinguishable.

(*b*) $\mathbf{C}_{4v}$. The elements of this group fall into classes as given by I, $2C_4$, C_2, $2\sigma_v$, $2\sigma_d$: there are therefore five symmetry species labelled A_1, A_2, B_1, B_2, E. The A, B, 1, and 2 labels have their usual meaning.

The character table is given in table 4.13.

(*c*) $\mathbf{C}_{\infty v}$. Like the $\mathbf{D}_{\infty h}$ point group and the full rotation-inversion group $\mathbf{K}_h$ the $\mathbf{C}_{\infty v}$ point group is an *infinite group* having an infinite number of elements, classes and symmetry species. The elements are I, an infinite number of σ_v planes, and an infinite number of elements C_∞^{ϕ}, $C_\infty^{2\phi}$, $C_\infty^{3\phi}$ . . . where ϕ is an arbitrary angle through which rotation about the C_∞ axis occurs. The element $C_\infty^{-\phi}$ belongs to the same class as C_∞^{ϕ}, $C_\infty^{-2\phi}$ to the same class as $C_\infty^{2\phi}$ etc. All the σ_v planes belong to one class.

The symmetry species of this point group are A_1, A_2, E_1, E_2, E_3, . . . E_∞ in a notation which is consistent with that used so far for other point groups. Unfortunately this notation is rarely used: instead the labels Σ^+, Σ^-, Π, Δ, Φ, . . . are usually used. The Greek letters in this notation signify the value of a quantum number associated with the component of the electronic orbital angular momentum along the internuclear axis. The quantum number has the values 0, 1, 2, 3 . . . corresponding to the symbols Σ, Π, Δ, Φ, . . . The superscript + and – on the Σ labels have the same meaning as the subscripts 1 and 2, respectively, on the alternative A labels.

The character table for the $\mathbf{C}_{\infty v}$ point group is given in table 4.16.

(*d*) $\mathbf{C}_3$. Molecules belonging to the $\mathbf{C}_3$ point group are very rare: an example would be $CH_3 . CF_3$ if the C–H and C–F bonds were at an angle of less than $\pi/3$ to each other as shown in figure 4.8. The $\mathbf{C}_3$ character table, however, is worthy of mention here since it is the simplest character table which illustrates the phenomenon of *separable degeneracy*. This character table is given in table 4.5.

Separable degeneracy arises in all point groups which have a higher than two-fold axis but which have no σ_v planes of symmetry, or C_2 axes perpendicular to the main axis. Examples of such point groups are $C_3, C_4, \ldots C_{3h}, C_{4h}, \ldots S_4, S_6, \ldots$ The reason for the degeneracy being called separable is that although an E vibration, for example, in such a point group is doubly degenerate, the normal co-ordinates Q_{ja} and Q_{jb} are each transformed by the operation C_n (where $n > 2$) into multiples of themselves

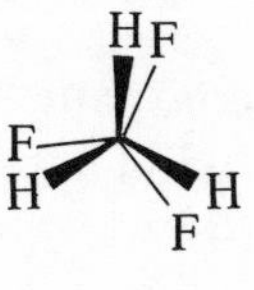

FIGURE 4.8
View down the C–C bond of $CH_3 . CF_3$ in which adjacent C–H and C–F bonds are at an angle of less than $\pi/3$ to each other.

and not into the linear combination given by equation 4.12. In order to prove this property of Q_{ja} and Q_{jb} it is convenient to use complex linear combinations Q_{j+} and Q_{j-} where

$$\begin{aligned} Q_{j+} &= Q_{ja} + iQ_{jb} \\ Q_{j-} &= Q_{ja} - iQ_{jb} \end{aligned} \tag{4.14}$$

where $i = \sqrt{-1}$. (It is a requirement of degenerate normal co-ordinates that linear combinations of them are equally valid normal co-ordinates.) Under the C_3 operation, for example, we get from equation 4.10

$$\begin{aligned} Q_{ja} &\xrightarrow{C_3} Q'_{ja} = Q_{ja}\cos 2\pi/3 + Q_{jb}\sin 2\pi/3 \\ Q_{jb} &\xrightarrow{C_3} Q'_{jb} = -Q_{ja}\sin 2\pi/3 + Q_{jb}\cos 2\pi/3 \end{aligned} \tag{4.15}$$

But

$$\begin{aligned} Q'_{j+} &= Q'_{ja} + iQ'_{jb} \\ Q'_{j-} &= Q'_{ja} - iQ'_{jb} \end{aligned} \tag{4.16}$$

and therefore

$$\begin{aligned} Q'_{j+} &= Q_{ja}\cos 2\pi/3 + Q_{jb}\sin 2\pi/3 + i(-Q_{ja}\sin 2\pi/3 + Q_{jb}\cos 2\pi/3) \\ Q'_{j-} &= Q_{ja}\cos 2\pi/3 + Q_{jb}\sin 2\pi/3 - i(-Q_{ja}\sin 2\pi/3 + Q_{jb}\cos 2\pi/3) \end{aligned}$$

Using $\cos 2\pi/3 \pm i \sin 2\pi/3 = \exp(\pm 2\pi i/3)$ and $\sin 2\pi/3 \pm i \cos 2\pi/3 = \pm i \exp(\mp 2\pi i/3)$ we get

$$\begin{aligned} Q'_{j+} &= (Q_{ja} + iQ_{jb}) \exp(-2\pi i/3) = Q_{j+} \exp(-2\pi i/3) \\ Q'_{j-} &= (Q_{ja} - iQ_{jb}) \exp(2\pi i/3) = Q_{j-} \exp(2\pi i/3) \end{aligned} \tag{4.17}$$

It can be seen that Q_{j+} is not transformed by the C_3 operation into a linear combination of Q_{j+} and Q_{j-} but is simply multiplied by an exponential factor. The same is true for Q_{j-}.

Because of the separable degeneracy, the characters of the two components are usually written separately as in the C_3 character table in table 4.5. There is, however, only one symmetry species label E which applies to both components.

It is a feature of point groups which have separable degeneracy that *all powers of the* C_n *(where* $n > 2$*) symmetry element belong to different classes.* For example in the $\mathbf{C_3}$ point group C_3 and C_3^2 belong to different classes, whereas in the $\mathbf{C_{3v}}$ point group they are in the same class. Also, in such point groups, each component of the doubly degenerate symmetry species has to be counted in order to make use of the rule that the number of symmetry species is equal to the number of classes. Thus the elements of the $\mathbf{C_3}$ point group fall into three classes, I, C_3, and C_3^2 and there are three symmetry species A and E (2 components).

(*e*) $\mathbf{D_3}$. This group does not show separable degeneracy since there are three C_2 axes perpendicular to the C_3 axis. The elements fall into three classes I, $2C_3$, $3C_2$ giving three symmetry species A_1, A_2, and E.

The character table is given in table 4.18.

(*f*) $\mathbf{C_{3h}}$. All the six elements of this point group, which exhibits separable degeneracy, fall into different classes I, C_3, C_3^2, σ_h, S_3, S_3^5 so there are six symmetry species A′, A″, E′ (two components), and E″ (two components). The single and double primes indicate species which are respectively symmetric and antisymmetric to σ_h.

The character table is given in table 4.23.

(*g*) $\mathbf{D_{2d}}$. The elements of the $\mathbf{D_{2d}}$ point group fall into the five classes I, $2S_4$, C_2, $2C'_2$, $2\sigma_d$ and the five symmetry species are A_1, A_2, B_1, B_2, and E. The A and B labels again imply symmetry or antisymmetry with respect to rotation about the main axis, in this case S_4.

The character table is given in table 4.27.

(*h*) $\mathbf{D_{3d}}$. The elements of the $\mathbf{D_{3d}}$ point group fall into the six classes I, $2C_3$, $3C_2$, i, $2S_6$, $3\sigma_d$ and the six symmetry species are A_{1g}, A_{2g}, E_g, A_{1u}, A_{2u}, E_u.

The character table is given in table 4.28.

(*i*) $\mathbf{D_{4d}}$. There are seven classes in the $\mathbf{D_{4d}}$ point group I, $2S_8$, $2C_4$, $2S_8{}^3$, C_2, $4C_2'$, $4\sigma_d$. The seven symmetry species are A_1, A_2, B_1, B_2, E_1, E_2, E_3. The subscript 1, 2, or 3 on the E species is the value of *l* required in equation 4.12.

The character table is given in table 4.29.

(*j*) $\mathbf{D_{5d}}$. The elements of the $\mathbf{D_{5d}}$ point group fall into eight classes I, $2C_5$, $2C_5{}^2$, $5C_2$, i, $2S_{10}{}^3$, $2S_{10}$, $5\sigma_d$ giving eight symmetry species A_{1g}, A_{2g}, E_{1g}, E_{2g}, A_{1u}, A_{2u}, E_{1u}, E_{2u}.

The character table is given in table 4.30.

(*k*) $\mathbf{D_{3h}}$. The elements of the $\mathbf{D_{3h}}$ point group fall into six classes I, $2C_3$, $3C_2$, σ_h, $2S_3$, $3\sigma_v$. The six symmetry species are A_1', A_2', E', A_1'', A_2'', E''.

The character table is given in table 4.33.

(*l*) $\mathbf{D_{4h}}$. There are ten classes in the $\mathbf{D_{4h}}$ point group I, $2C_4$, C_2, $2C_2'$, $2C_2''$, i, $2S_4$, σ_h, $2\sigma_v$, $2\sigma_d$ and the ten symmetry species are A_{1g}, A_{2g}, B_{1g}, B_{2g}, E_g, A_{1u}, A_{2u}, B_{1u}, B_{2u}, E_u.

The character table is given in table 4.34.

(*m*) $\mathbf{D_{5h}}$. The elements of the $\mathbf{D_{5h}}$ point group fall into eight classes I, $2C_5$, $2C_5{}^2$, $5C_2$, σ_h, $2S_5$, $2S_5{}^3$, $5\sigma_v$. The eight symmetry species are A_1', A_2', E_1', E_2', A_1'', A_2'', E_1'', E_2''.

The character table is given in table 4.35.

(*n*) $\mathbf{D_{6h}}$. The elements of the $\mathbf{D_{6h}}$ point group fall into twelve classes I, $2C_6$, $2C_3$, C_2, $3C_2'$, $3C_2''$, i, $2S_3$, $2S_6$, σ_h, $3\sigma_d$, $3\sigma_v$. The twelve symmetry species are A_{1g}, A_{2g}, B_{1g}, B_{2g}, E_{1g}, E_{2g}, A_{1u}, A_{2u}, B_{1u}, B_{2u}, E_{1u}, E_{2u}.

The character table is given in table 4.36.

(*o*) $\mathbf{D_{\infty h}}$: The $\mathbf{D_{\infty h}}$ point group has an infinite number of elements, classes and symmetry species. The symmetry elements fall into classes I, $\infty\sigma_v$, i, ∞C_2, $2C_\infty{}^\phi$, $2C_\infty{}^{2\phi}$ . . . , $2S_\infty{}^\phi$, $2S_\infty{}^{2\phi}$ . . . The symmetry species of this point group are A_{1g}, A_{2g}, E_{1g}, E_{2g}, E_{3g}, . . ., A_{2u}, A_{1u}, E_{1u}, E_{2u}, E_{3u}, . . . but, in an analogous way to the $C_{\infty v}$ point group, these are usually replaced by Σ_g^+, Σ_g^-, Π_g, Δ_g, Φ_g, . . ., Σ_u^+, Σ_u^-, Π_u, Δ_u, Φ_u, . . . respectively.

The character table for the $D_{\infty h}$ point group is given in table 4.37.

The case of a doubly degenerate π_u vibration in a linear triatomic molecule belonging to the $D_{\infty h}$ point group (for example CO_2) provides an example for which it is easy to derive the character of the symmetry species with respect to the C_∞^ϕ operation.

CO_2 has $3N - 5 = 4$ normal vibrations labelled $\nu_1, \nu_{2a}, \nu_{2b}, \nu_3$ and illustrated in figure 4.9. Unlike the case of the degenerate vibrations ν_{3a}

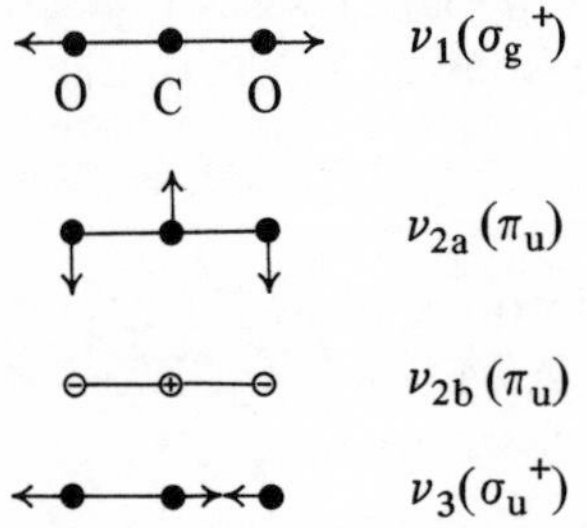

FIGURE 4.9
The normal vibrations of CO_2

and ν_{3b} in ammonia (figure 4.5), it is obvious that ν_{2a} and ν_{2b} in CO_2 are energetically equivalent and therefore degenerate. The symmetry species is π_u. The motions of the three atoms in the ν_{2a} normal mode are the same as, but perpendicular to, those in ν_{2b}. In general these motions can be represented by vectors each of length r_k^a for nucleus k in the vibration ν_{2a} and r_k^b for the same nucleus in ν_{2b}. These vectors are perpendicular to each other and are at an angle α, say, to the y- and x-axes respectively, as illustrated in figure 4.10 in which the C_∞ axis is the axis perpendicular to the plane of the figure. The component along the x-axis of, for example, the vector of magnitude r_k^a is x_k^a and along the y-axis is y_k^a. If the vectors are rotated clockwise by an angle ϕ about the C_∞ axis the components of the vector of magnitude r_k^a along the cartesian axes become $(x_k^a)'$ and $(y_k^a)'$ where

$$\begin{aligned}(x_k^a)' &= r_k^a \sin(\alpha + \phi) = x_k^a \cos\phi + y_k^a \sin\phi \\ (y_k^a)' &= r_k^a \cos(\alpha + \phi) = y_k^a \cos\phi - x_k^a \sin\phi\end{aligned} \tag{4.18}$$

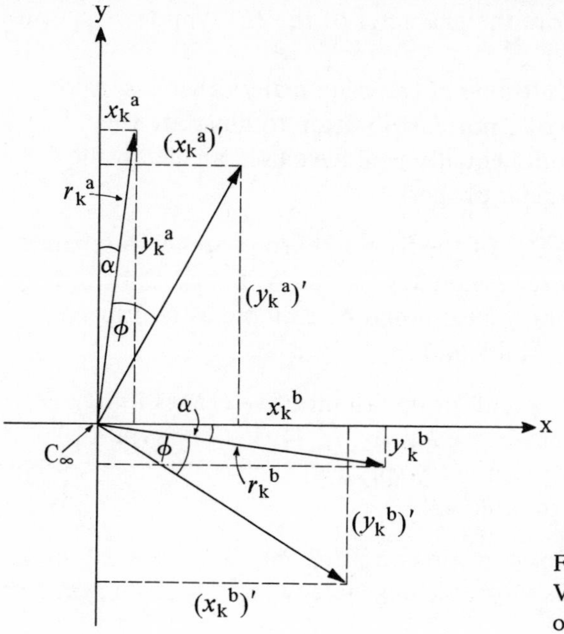

FIGURE 4.10
Vector diagram for the C_∞^ϕ operation on ν_{2a} and ν_{2b} in CO_2

But $y_k{}^a = x_k{}^b$ and $x_k{}^a = -y_k{}^b$ therefore

$$\begin{aligned}(x_k{}^a)' &= x_k{}^a \cos\phi + x_k{}^b \sin\phi \\ (y_k{}^a)' &= y_k{}^a \cos\phi + y_k{}^b \sin\phi\end{aligned} \tag{4.19}$$

Similarly for ν_{2b}

$$\begin{aligned}(x_k{}^b)' &= -x_k{}^a \sin\phi + x_k{}^b \cos\phi \\ (y_k{}^b)' &= -y_k{}^a \sin\phi + y_k{}^b \cos\phi\end{aligned} \tag{4.20}$$

Since equations 4.19 and 4.20 hold for any nucleus k moving in each of the degenerate normal modes

$$\begin{aligned}Q'_{2a} &= Q_{2a} \cos\phi + Q_{2b} \sin\phi \\ Q'_{2b} &= -Q_{2a} \sin\phi + Q_{2b} \cos\phi\end{aligned} \tag{4.21}$$

The matrix representing this transformation is

$$\begin{pmatrix} \cos\phi & \sin\phi \\ -\sin\phi & \cos\phi \end{pmatrix} \tag{4.22}$$

and the trace, $2 \cos \phi$, is therefore the character of the Π_u symmetry species under the symmetry operation C_∞^ϕ.

It must be remembered that although here, as in many other cases, we have used a particular example of a normal vibration to illustrate the derivation of a character, we could equally well have used an electronic wave function or another molecular property.

(*p*) $\mathbf{S_4}$. The elements I, S_4, C_2, S_4^3 of the $\mathbf{S_4}$ point group are all in separate classes. There are really only three symmetry species A, B, and E, but the E species shows separable degeneracy and should be counted as two species.

The character table is given in table 4.38.

(*q*) $\mathbf{T_d}$. The elements of the $\mathbf{T_d}$ point group fall into five classes I, $8C_3$, $3C_2$, $6S_4$, $6\sigma_d$. The five symmetry species are A_1, A_2, E, T_1, T_2. The symbol T indicates a threefold degenerate symmetry species.†

The character table is given in table 4.41.

(*r*) **T**. The elements of the **T** point group comprise only four classes I, $4C_3$, $4C_3^2$, $3C_2$. The point group has separable degeneracy and the three symmetry species are A, E, T.

The character table is given in table 4.42.

(*s*) $\mathbf{O_h}$. The elements of the $\mathbf{O_h}$ point group fall into ten classes I, $8C_3$, $6C_2$, $6C_4$, $3C_2 (= C_4^2)$, i, $6S_4$, $8S_6$, $3\sigma_h$, $6\sigma_d$ and the ten symmetry species are A_{1g}, A_{2g}, E_g, T_{1g}, T_{2g}, A_{1u}, A_{2u}, E_u, T_{1u}, T_{2u}.

The character table is given in table 4.43.

(*t*) **O**. The elements of this group fall into five classes I, $8C_3$, $6C_2$, $6C_4$, $3C_2'(= C_4^2)$ and the five symmetry species are A_1, A_2, E, T_1, T_2.

The character table is given in table 4.44.

(*u*) $\mathbf{C_4}$, $\mathbf{C_5}$, $\mathbf{C_6}$, $\mathbf{C_7}$, $\mathbf{C_8}$, $\mathbf{C_{4h}}$, $\mathbf{C_{5h}}$, $\mathbf{C_{6h}}$, $\mathbf{S_6}$, $\mathbf{S_8}$, $\mathbf{D_4}$, $\mathbf{D_5}$, $\mathbf{D_6}$, $\mathbf{C_{5v}}$, $\mathbf{C_{6v}}$, $\mathbf{D_{6d}}$. These degenerate point groups are all extremely rare among molecules, but for the sake of some degree of completeness their character tables are given with the others. The index on p. 46 to the tables can be used to find the one required.

† F is often used instead of T: Mulliken recommends either but T is preferable in some cases, especially in ligand field theory and crystal field theory (section 6.5), in which f atomic orbitals which are *sevenfold* degenerate, are important.

(v) $\mathbf{K_h}$. The full rotation-inversion point group $\mathbf{K_h}$ is the point group to which all atoms belong by virtue of their spherical symmetry. There are an infinite number of symmetry elements (I, an infinite number of rotation axes and planes of symmetry, and a centre of symmetry), classes and symmetry species in this point group. The S, P, D, F . . . labels for the symmetry species have a curious historical origin, but now imply that an atom in an electronic state to which one of these labels is given has the orbital angular momentum quantum number $L = 0, 1, 2, 3$ . . . respectively.

The character table is given in table 4.45.

4.3 Multiplication of symmetry species

In many of the applications of the theory of molecular symmetry it will be necessary to be able to multiply together any number of symmetry species in a particular point group. For example if a molecule is in an electronic state with a wave function of symmetry species A and at the same time vibrating with one quantum of a vibration of symmetry species b then the *vibronic* wave function ψ_{ev} ($= \psi_e \times \psi_v$) has the species given by A x b. Also if a molecule is vibrating in two modes at the same time with one quantum of a vibration of species a and one quantum of species b then the symmetry species of the total vibrational wave function is given by a x b.

4.3.1 MULTIPLICATION OF TWO NON-DEGENERATE SYMMETRY SPECIES

In non-degenerate point groups multiplication of species is very easy. To obtain the characters of the product of two species with respect to a symmetry operation one simply multiplies the characters of the two species with respect to that symmetry operation. For example, table 4.46 illustrates

TABLE 4.46
The product $A_2 \times B_1$ in the $\mathbf{C_{2v}}$ point group

$\mathbf{C_{2v}}$	I	C_2	$\sigma_v(xz)$	$\sigma_v'(yz)$
A_2	1	1	−1	−1
B_1	1	−1	1	−1
$A_2 \times B_1 (=B_2)$	1	−1	−1	1

the result that in the $\mathbf{C}_{2v}$ point group $A_2 \times B_1 = B_2$. Table 4.47 gives a complete multiplication table for the species of the $\mathbf{C}_{2v}$ point group. The results can be verified using the character table in table 4.11. It is obviously a

TABLE 4.47

Symmetry species multiplication table for the $\mathbf{C}_{2v}$ point group

	A_1	A_2	B_1	B_2
A_1	A_1	A_2	B_1	B_2
A_2	A_2	A_1	B_2	B_1
B_1	B_1	B_2	A_1	A_2
B_2	B_2	B_1	A_2	A_1

general result that the square of any non-degenerate symmetry species must give the totally symmetric species; for example, in the $\mathbf{C}_{2v}$ point group $B_1 \times B_1 = A_1$. It is also a general result that any symmetry species (non-degenerate or degenerate) is unchanged on multiplication by the totally symmetric species; for example, in the $\mathbf{C}_{2v}$ point group, $A_2 \times A_1 = A_2$.

Similarly, any number of species can be multiplied; for example, in the $\mathbf{C}_{2v}$ point group, $A_2 \times B_1 \times B_2 = A_2 \times A_2 = A_1$.

Multiplication of non-degenerate symmetry species is so easy that it is not necessary to present multiplication tables similar to that in table 4.47 for all point groups. Nor is it advisable to attempt to memorize the results of multiplication, but it can be useful and time-saving to remember a few general rules in addition to those already mentioned.

(*a*) If the species have g or u subscripts then, in any point group,

$$g \times g = g; \quad u \times u = g; \quad g \times u = u \tag{4.23}$$

(*b*) If the species have single or double primes then, in any point group,

$$(') \times (') = ('); \quad ('') \times ('') = ('); \quad ('') \times (') = ('') \tag{4.24}$$

(*c*) In point groups where subscripts to A and B species cannot be greater than 2

$$\text{(i)} \quad A \times A = A; \quad B \times B = A; \quad A \times B = B \tag{4.25}$$

and for the subscripts

(ii) $1 \times 1 = 1$; $2 \times 2 = 1$; $1 \times 2 = 2$ (4.26)

(*d*) In the $\mathbf{D}_{2h}$ point group the subscripts 1, 2, and 3 multiply in a cyclic manner

$$1 \times 2 = 3; \quad 2 \times 3 = 1; \quad 3 \times 1 = 2 \tag{4.27}$$

4.3.2 MULTIPLICATION OF ONE NON-DEGENERATE AND ONE DEGENERATE SYMMETRY SPECIES

If one of the symmetry species to be multiplied is degenerate then the resulting species is obtained by the same method as described in section 4.3.1 and is always a degenerate species. Table 4.48 illustrates that, in the $\mathbf{C}_{3v}$ point group, $A_2 \times E = E$ and that, in the **O** point group, $A_2 \times T_2 = T_1$.

TABLE 4.48
Two examples of multiplication of a non-degenerate and a degenerate symmetry species

C_{3v}	I	$2C_3$	$3\sigma_v$
A_2	1	1	−1
E	2	−1	0
$A_2 \times E(=E)$	2	−1	0

O	I	$8C_3$	$6C_2$	$6C_4$	$3C_2'(=3C_4^2)$
A_2	1	1	−1	−1	1
T_2	3	0	1	−1	−1
$A_2 \times T_2(=T_1)$	3	0	−1	1	−1

4.3.3 MULTIPLICATION OF TWO DEGENERATE SYMMETRY SPECIES

This type of multiplication is most easily discussed using normal modes of vibration as examples.

It is an important feature of the multiplication of degenerate symmetry species that the result of multiplying two identical symmetry species depends on whether, in the vibrational example, this represents a molecule vibrating with two quanta of the same vibration or with one quantum of each of two different vibrations with the same symmetry species. For example the symmetry species of the vibrational states resulting when ammonia vibrates with two quanta of ν_3 (figure 4.5) are given by

$$(e)^2 = A_1 + E \tag{4.28}$$

whereas the species of the states resulting when it vibrates with one quantum of ν_3 and one quantum of ν_4 (figure 4.5) are given by

$$e \times e = A_1 + A_2 + E \tag{4.29}$$

(Note the way in which these two cases are distinguished by using $(e)^2$ or $e \times e$. Note also the use of lower case letters for the symmetry species of a normal vibration and upper case letters for the species of vibrational states.)

The result of the product $e \times e$ can be separated into two parts, $A_1 + E$ which is called the *symmetric* part and A_2 which is the *antisymmetric* part. $(e)^2$ contains the symmetric part of $e \times e$ and this rule is quite general. When we come to derive electronic states which arise from two electrons in degenerate orbitals (section 6.6) we shall see that we always use products of the type $e \times e$ and that the symmetric part of the product gives singlet states and the antisymmetric part gives triplet states.

How do we obtain the results of equations 4.28 and 4.29?

In order to obtain the product $e \times e$ the character of e with respect to each symmetry operation is squared and the resulting characters always represent

TABLE 4.49

The product $e \times e$ in the $\mathbf{C}_{3v}$ point group

C_{3v}	I	$2C_3$	$3\sigma_v$
A_1	1	1	1
A_2	1	1	−1
E	2	−1	0
$e \times e$	4	1	0

the sum of the characters of a unique combination of symmetry species, in this case $A_1 + A_2 + E$ as shown in table 4.49. If χ represents a character then, for example,

$$\chi_E(C_3) \times \chi_E(C_3) = \chi_{A_1}(C_3) + \chi_{A_2}(C_3) + \chi_E(C_3) \tag{4.30}$$

for the C_3 operation.

Table 4.50 illustrates that in the $\mathbf{D}_{6h}$ point group $e_{2u} \times e_{2u} = A_{1g} + A_{2g} + E_{2g}$ and shows also that in carrying out multiplication of symmetry species one need consider only a set of generating elements in the group and not all the elements.

In general the expression

$$\chi_C(k) \times \chi_D(k) = \chi_F(k) + \chi_G(k) + \cdots \tag{4.31}$$

where k is any operation, enables one to determine uniquely the species F + G + ... resulting from the multiplication of two degenerate symmetry species C and D.

TABLE 4.50
The products $e_{2u} \times e_{2u}$ and $e_{1g} \times e_{2u}$ in the $\mathbf{D_{6h}}$ point group

D_{6h}	$2C_3$	C_2	$3C_2'$	i	σ_h	$3\sigma_v$
A_{1g}	1	1	1	1	1	1
A_{2g}	1	1	−1	1	1	−1
E_{2g}	−1	2	0	2	2	0
$e_{2u} \times e_{2u}$	1	4	0	4	4	0
B_{1u}	1	−1	1	−1	1	1
B_{2u}	1	−1	−1	−1	1	−1
E_{1u}	−1	−2	0	−2	2	0
$e_{1g} \times e_{2u}$	1	−4	0	−4	4	0

So far we have considered only examples where C = D but equation 4.31 is quite general and table 4.50 illustrates that, in the $\mathbf{D_{6h}}$ point group, $e_{1g} \times e_{2u} = B_{1u} + B_{2u} + E_{1u}$.

In the cases of degenerate symmetry species where the degeneracy is separable, the sum of the characters, with respect to any symmetry operation, of the product of two degenerate symmetry species is obtained by multiplying each of the two characters of the degenerate species in all possible combinations (four combinations for two doubly degenerate species). The examples in table 4.51 show that in the $\mathbf{C_7}$ point, for which the generating element is C_7, $e_1 \times e_1 = 2A + E_2$ and $e_2 \times e_3 = E_1 + E_2$.

TABLE 4.51
The products $e_1 \times e_1$ and $e_2 \times e_3$ in the $\mathbf{C}_7$ point group

C_7	C_7
A	1
E_2	ϵ^2 ϵ^{2*}
$e_1 \times e_1$	ϵ^2 ϵ^{2*} 1 1
E_1	ϵ ϵ^*
E_2	ϵ^2 ϵ^{2*}
$e_2 \times e_3$	ϵ ϵ^* ϵ^2 ϵ^{2*}

Table 4.52 gives the states resulting from all possible pairwise combinations of one quantum of different degenerate vibrations for all the degenerate point groups included in tables 4.1 to 4.44.

In order to obtain the $(c)^2$ product, where c is any doubly degenerate species, one first obtains the product c x c using the method already discussed. The result will always contain two species with an A label. *In order to go from the states resulting from c x c to those resulting from $(c)^2$ one of the A species is eliminated: where possible this is the one which is nontotally symmetric.* For example in the $\mathbf{C}_{3v}$ point group A_2 is eliminated from the $A_1 + A_2 + E$ states of e x e (equation 4.29) to give the states $A_1 + E$ of $(e)^2$ (equation 4.28). In the $\mathbf{C}_4$ point group e x e = 2A + 2B but $(e)^2$ = A + 2B.

TABLE 4.52
Symmetry species of vibrational states resulting from one quantum of each of two different degenerate vibrations

Point group	*Symmetry species of two quanta vibrational states*
$\mathbf{C_3}$	$e \times e = 2A + E$
$\mathbf{C_4}$	$e \times e = 2A + 2B$
$\mathbf{C_5}$	$e_1 \times e_1 = 2A + E_2,\ e_2 \times e_2 = 2A + E_1,\ e_1 \times e_2 = E_1 + E_2$
$\mathbf{C_6}$	$e_1 \times e_1 = 2A + E_2,\ e_2 \times e_2 = 2A + E_2,\ e_1 \times e_2 = 2B + E_1$
$\mathbf{C_7}$	$e_1 \times e_1 = 2A + E_2,\ e_2 \times e_2 = 2A + E_3,\ e_3 \times e_3 = 2A + E_1,\ e_1 \times e_2 = E_1 + E_3,$ $e_1 \times e_3 = E_2 + E_3,\ e_2 \times e_3 = E_1 + E_2$
$\mathbf{C_8}$	$e_1 \times e_1 = 2A + E_2,\ e_2 \times e_2 = 2A + 2B,\ e_3 \times e_3 = 2A + E_2,\ e_1 \times e_2 = E_1 + E_3,$ $e_1 \times e_3 = 2B + E_2,\ e_2 \times e_3 = E_1 + E_3$
$\mathbf{C_{3v}}$	$e \times e = A_1 + A_2 + E$
$\mathbf{C_{4v}}$	$e \times e = A_1 + A_2 + B_1 + B_2$
$\mathbf{C_{5v}}$	$e_1 \times e_1 = A_1 + A_2 + E_2,\ e_2 \times e_2 = A_1 + A_2 + E_1,\ e_1 \times e_2 = E_1 + E_2$
$\mathbf{C_{6v}}$	$e_1 \times e_1 = A_1 + A_2 + E_2,\ e_2 \times e_2 = A_1 + A_2 + E_2,\ e_1 \times e_2 = B_1 + B_2 + E_1$
$\mathbf{C_{\infty v}}$	$\pi \times \pi = \Sigma^+ + \Sigma^- + \Delta,\ \delta \times \delta = \Sigma^+ + \Sigma^- + \Gamma,\ \pi \times \delta = \Pi + \Phi$
$\mathbf{D_3}$	$e \times e = A_1 + A_2 + E$
$\mathbf{D_4}$	$e \times e = A_1 + A_2 + B_1 + B_2$
$\mathbf{D_5}$	$e_1 \times e_1 = A_1 + A_2 + E_2,\ e_2 \times e_2 = A_1 + A_2 + E_1,\ e_1 \times e_2 = E_1 + E_2$
$\mathbf{D_6}$	$e_1 \times e_1 = A_1 + A_2 + E_2,\ e_2 \times e_2 = A_1 + A_2 + E_2,\ e_1 \times e_2 = B_1 + B_2 + E_1$
$\mathbf{C_{3h}}$	$e' \times e' = 2A' + E',\ e'' \times e'' = 2A' + E',\ e' \times e'' = 2A'' + E''$
$\mathbf{C_{4h}}$	$e_g \times e_g = 2A_g + 2B_g,\ e_u \times e_u = 2A_g + 2B_g,\ e_g \times e_u = 2A_u + 2B_u$
$\mathbf{C_{5h}}$	$e_1' \times e_1' = 2A' + E_2',\ e_2' \times e_2' = 2A' + E_1',\ e_1'' \times e_1'' = 2A' + E_2',\ e_2'' \times e_2'' = 2A' + E_1',$ $e_1' \times e_1'' = 2A + E_2'',\ e_1' \times e_2'' = E_1'' + E_2'',\ e_1' \times e_2' = E_1' + E_2',\ e_1' \times e_2'' = E_1'' + E_2'',$ $e_1'' \times e_2' = E_1'' + E_2'',\ e_1'' \times e_2'' = E_1' + E_2'$
$\mathbf{C_{6h}}$	$e_{1g} \times e_{1g} = 2A_g + E_{2g},\ e_{2g} \times e_{2g} = 2A_g + E_{2g},\ e_{1u} \times e_{1u} = 2A_g + E_{2g},$ $e_{2u} \times e_{2u} = 2A_g + E_{2g},\ e_{1g} \times e_{1u} = 2A_u + E_{2u},\ e_{1g} \times e_{2g} = 2B_g + E_{1g},$ $e_{1g} \times e_{2u} = 2B_u + E_{1u},\ e_{1u} \times e_{2g} = 2B_u + E_{1u},\ e_{1u} \times e_{2u} = 2B_g + 2E_{1g},$ $e_{2g} \times e_{2u} = 2A_u + E_{2u}$

TABLE 4.52 continued

Point group	*Symmetry species of two quanta vibrational states*
$\mathbf{D_{2d}}$	$e \times e = A_1 + A_2 + B_1 + B_2$
$\mathbf{D_{3d}}$	$e_g \times e_g = A_{1g} + A_{2g} + E_g$, $e_u \times e_u = A_{1g} + A_{2g} + E_g$, $e_g \times e_u = A_{1u} + A_{2u} + E_u$
$\mathbf{D_{4d}}$	$e_1 \times e_1 = A_1 + A_2 + E_2$, $e_2 \times e_2 = A_1 + A_2 + B_1 + B_2$, $e_3 \times e_3 = A_1 + A_2 + E_2$, $e_1 \times e_2 = E_1 + E_3$, $e_1 \times e_3 = B_1 + B_2 + E_2$, $e_2 \times e_3 = E_1 + E_3$
$\mathbf{D_{5d}}$	$e_{1g} \times e_{1g} = A_{1g} + A_{2g} + E_{2g}$, $e_{2g} \times e_{2g} = A_{1g} + A_{2g} + E_{1g}$, $e_{1u} \times e_{1u} = A_{1g} + A_{2g} + E_{2g}$, $e_{2u} \times e_{2u} = A_{1g} + A_{2g} + E_{1g}$, $e_{1g} \times e_{2g} = E_{1g} + E_{2g}$, $e_{1g} \times e_{1u} = A_{1u} + A_{2u} + E_{2u}$, $e_{1u} \times e_{2u} = E_{1g} + E_{2g}$, $e_{1g} \times e_{2u} = E_{1u} + E_{2u}$, $e_{2g} \times e_{1u} = E_{1u} + E_{2u}$, $e_{2g} \times e_{2u} = A_{1u} + A_{2u} + E_{1u}$
$\mathbf{D_{6d}}$	$e_1 \times e_1 = A_1 + A_2 + E_2$, $e_2 \times e_2 = A_1 + A_2 + E_4$, $e_3 \times e_3 = A_1 + A_2 + B_1 + B_2$, $e_4 \times e_4 = A_1 + A_2 + E_4$, $e_5 \times e_5 = A_1 + A_2 + E_2$, $e_1 \times e_2 = E_1 + E_3$, $e_1 \times e_3 = E_2 + E_4$, $e_1 \times e_4 = E_3 + E_5$, $e_1 \times e_5 = B_1 + B_2 + E_4$, $e_2 \times e_3 = E_1 + E_5$, $e_2 \times e_4 = B_1 + B_2 + E_2$, $e_2 \times e_5 = E_3 + E_5$, $e_3 \times e_4 = E_1 + E_5$, $e_3 \times e_5 = E_2 + E_4$, $e_4 \times e_5 = E_1 + E_3$
$\mathbf{D_{3h}}$	$e' \times e' = A_1' + A_2' + E'$, $e'' \times e'' = A_1' + A_2' + E'$, $e' \times e'' = A_1'' + A_2'' + E''$
$\mathbf{D_{4h}}$	$e_g \times e_g = A_{1g} + A_{2g} + B_{1g} + B_{2g}$, $e_u \times e_u = A_{1g} + A_{2g} + B_{1g} + B_{2g}$, $e_g \times e_u = A_{1u} + A_{2u} + B_{1u} + B_{2u}$
$\mathbf{D_{5h}}$	$e_1' \times e_1' = A_1' + A_2' + E_2'$, $e_1'' \times e_1'' = A_1' + A_2' + E_2'$, $e_2' \times e_2' = A_1' + A_2' + E_1'$, $e_2'' \times e_2'' = A_1' + A_2' + E_1'$, $e_1' \times e_1'' = A_1'' + A_2'' + E_2''$, $e_1' \times e_2' = E_1' + E_2'$, $e_1' \times e_2'' = E_1'' + E_2''$, $e_1'' \times e_2' = E_1'' + E_2''$, $e_1'' \times e_2'' = E_1' + E_2'$, $e_2' \times e_2'' = A_1'' + A_2'' + E_1''$
$\mathbf{D_{6h}}$	$e_{1g} \times e_{1g} = A_{1g} + A_{2g} + E_{2g}$, $e_{2g} \times e_{2g} = A_{1g} + A_{2g} + E_{2g}$, $e_{1u} \times e_{1u} = A_{1g} + A_{2g} + E_{2g}$, $e_{2u} \times e_{2u} = A_{1g} + A_{2g} + E_{2g}$, $e_{1g} \times e_{1u} = A_{1u} + A_{2u} + E_{2u}$, $e_{2g} \times e_{2u} = A_{1u} + A_{2u} + E_{2u}$, $e_{1g} \times e_{2g} = B_{1g} + B_{2g} + E_{1g}$, $e_{1g} \times e_{2u} = B_{1u} + B_{2u} + E_{1u}$, $e_{1u} \times e_{2g} = B_{1u} + B_{2u} + E_{1u}$, $e_{1u} \times e_{2u} = B_{1g} + B_{2g} + E_{1g}$
$\mathbf{D_{\infty h}}$	$\pi_g \times \pi_g = \Sigma_g^+ + \Sigma_g^- + \Delta_g$, $\delta_g \times \delta_g = \Sigma_g^+ + \Sigma_g^- + \Gamma_g$, $\pi_u \times \pi_u = \Sigma_g^+ + \Sigma_g^- + \Delta_g$, $\delta_u \times \delta_u = \Sigma_g^+ + \Sigma_g^- + \Gamma_g$, $\pi_g \times \pi_u = \Sigma_u^+ + \Sigma_u^- + \Delta_u$, $\delta_g \times \delta_u = \Sigma_u^+ + \Sigma_u^- + \Gamma_u$, $\pi_g \times \delta_g = \Pi_g + \Phi_g$, $\pi_u \times \delta_u = \Pi_g + \Phi_g$, $\pi_g \times \delta_u = \Pi_u + \Phi_u$, $\pi_u \times \delta_g = \Pi_u \times \Phi_u$

TABLE 4.52 continued

Point group	*Symmetry species of two quanta vibrational states*
S_4	$e \times e = 2A + 2B$
S_6	$e_g \times e_g = 2A_g + E_g$, $e_u \times e_u = 2A_g + E_g$, $e_g \times e_u = 2A_u + E_u$
S_8	$e_1 \times e_1 = 2A + E_2$, $e_2 \times e_2 = 2A + 2B$, $e_3 \times e_3 = 2A + E_2$, $e_1 \times e_2 = E_1 + E_3$, $e_1 \times e_3 = 2B + E_2$, $e_2 \times e_3 = E_1 + E_3$
T_d	$e \times e = A_1 + A_2 + E$, $t_1 \times t_1 = A_1 + E + T_1 + T_2$, $t_2 \times t_2 = A_1 + E + T_1 + T_2$, $e \times t_1 = T_1 + T_2$, $e \times t_2 = T_1 + T_2$, $t_1 \times t_2 = A_2 + E + T_1 + T_2$
T	$e \times e = 2A + E$, $t \times t = A + E + 2T$, $e \times t = 2T$
O_h	$e_g \times e_g = A_{1g} + A_{2g} + E_g$, $e_u \times e_u = A_{1g} + A_{2g} + E_g$, $e_g \times e_u = A_{1u} + A_{2u} + E_u$, $t_{1g} \times t_{1g} = A_{1g} + E_g + T_{1g} + T_{2g}$, $t_{1u} \times t_{1u} = A_{1g} + E_g + T_{1g} + T_{2g}$, $t_{1g} \times t_{1u} = A_{1u} + E_u + T_{1u} + T_{2u}$, $t_{2g} \times t_{2g} = A_{1g} + E_g + T_{1g} + T_{2g}$, $t_{2u} \times t_{2u} = A_{1g} + E_g + T_{1g} + T_{2g}$, $t_{2g} \times t_{2u} = A_{1u} + E_u + T_{1u} + T_{2u}$, $e_g \times t_{1g} = T_{1g} + T_{2g}$, $e_u \times t_{1u} = T_{1g} + T_{2g}$, $e_g \times t_{1u} = T_{1u} + T_{2u}$, $e_u \times t_{1g} = T_{1u} + T_{2u}$, $e_g \times t_{2g} = T_{1g} + T_{2g}$, $e_u \times t_{2u} = T_{1g} + T_{2g}$, $e_g \times t_{2u} = T_{1u} + T_{2u}$, $e_u \times t_{2g} = T_{1u} + T_{2u}$, $t_{1g} \times t_{2g} = A_{2g} + E_g + T_{1g} + T_{2g}$, $t_{1u} \times t_{2u} = A_{2g} + E_g + T_{1g} + T_{2g}$, $t_{1u} \times t_{2g} = A_{2u} + E_u + T_{1u} + T_{2u}$, $t_{1g} \times t_{2u} = A_{2u} + E_u + T_{1u} + T_{2u}$
O	$e \times e = A_1 + A_2 + E$, $t_1 \times t_1 = A_1 + E + T_1 + T_2$, $t_2 \times t_2 = A_1 + E + T_1 + T_2$, $e \times t_1 = T_1 + T_2$, $e \times t_2 = T_1 + T_2$, $t_1 \times t_2 = A_2 + E + T_1 + T_2$

The rule does not apply to obtaining the states resulting from two quanta of a vibration with a degeneracy greater than two-fold.

Table 4.53 gives the states resulting from all possible symmetric squares of degenerate symmetry species for all the degenerate point groups included in tables 4.1 to 4.44.

4.4 Symmetry species of rotations and translations

The three *principal axes* of a molecule are mutually perpendicular and intersect at the centre of gravity. One of the axes is the axis of maximum moment

TABLE 4.53
Symmetry species of vibrational states resulting from two quanta of one degenerate vibration

Point group	*Symmetry species of two quanta vibrational states*
$\mathbf{C_3}$	$(e)^2 = A + E$
$\mathbf{C_4}$	$(e)^2 = A + 2B$
$\mathbf{C_5}$	$(e_1)^2 = A + E_2$, $(e_2)^2 = A + E_1$
$\mathbf{C_6}$	$(e_1)^2 = A + E_2$, $(e_2)^2 = A + E_2$
$\mathbf{C_7}$	$(e_1)^2 = A + E_2$, $(e_2)^2 = A + E_3$, $(e_3)^2 = A + E_1$
$\mathbf{C_8}$	$(e_1)^2 = A + E_2$ $(e_2)^2 = A + 2B$, $(e_3)^2 = A + E_2$
$\mathbf{C_{3v}}$	$(e)^2 = A_1 + E$
$\mathbf{C_{4v}}$	$(e)^2 = A_1 + B_1 + B_2$
$\mathbf{C_{5v}}$	$(e_1)^2 = A_1 + E_2$, $(e_2)^2 = A_1 + E_1$
$\mathbf{C_{6v}}$	$(e_1)^2 = A_1 + E_2$, $(e_2)^2 = A_1 + E_2$
$\mathbf{C_{\infty v}}$	$(\pi)^2 = \Sigma^+ + \Delta$, $(\delta)^2 = \Sigma^+ + \Gamma$
$\mathbf{D_3}$	$(e)^2 = A_1 + E$
$\mathbf{D_4}$	$(e)^2 = A_1 + B_1 + B_2$
$\mathbf{D_5}$	$(e_1)^2 = A_1 + E_2$, $(e_2)^2 = A_1 + E_1$
$\mathbf{D_6}$	$(e_1)^2 = A_1 + E_2$, $(e_2)^2 = A_1 + E_2$
$\mathbf{C_{3h}}$	$(e')^2 = A' + E'$, $(e'')^2 = A' + E'$
$\mathbf{C_{4h}}$	$(e_g)^2 = A_g + 2B_g$, $(e_u)^2 = A_g + 2B_g$
$\mathbf{C_{5h}}$	$(e_1')^2 = A' + E_2'$, $(e_2')^2 = A' + E_1'$, $(e_1'')^2 = A' + E_2'$, $(e_2'')^2 = A' + E_1'$
$\mathbf{C_{6h}}$	$(e_{1g})^2 = A_g + E_{2g}$, $(e_{2g})^2 = A_g + E_{2g}$, $(e_{1u})^2 = A_g + E_{2g}$, $(e_{2u})^2 = A_g + E_{2g}$
$\mathbf{D_{2d}}$	$(e)^2 = A_1 + B_1 + B_2$
$\mathbf{D_{3d}}$	$(e_g)^2 = A_{1g} + E_g$, $(e_u)^2 = A_{1g} + E_g$
$\mathbf{D_{4d}}$	$(e_1)^2 = A_1 + E_2$, $(e_2)^2 = A_1 + B_1 + B_2$, $(e_3)^2 = A_1 + E_2$
$\mathbf{D_{5d}}$	$(e_{1g})^2 = A_{1g} + E_{2g}$, $(e_{2g})^2 = A_{1g} + E_{1g}$, $(e_{1u})^2 = A_{1g} + E_{2g}$, $(e_{2u})^2 = A_{1g} + E_{1g}$
$\mathbf{D_{6d}}$	$(e_1)^2 = A_1 + E_2$, $(e_2)^2 = A_1 + E_4$, $(e_3)^2 = A_1 + B_1 + B_2$, $(e_4)^2 = A_1 + E_4$, $(e_5)^2 = A_1 + E_2$
$\mathbf{D_{3h}}$	$(e')^2 = A_1' + E'$, $(e'')^2 = A_1' + E'$

TABLE 4.53 continued

Point group	*Symmetry species of two quanta vibrational states*
$\mathbf{D_{4h}}$	$(e_g)^2 = A_{1g} + B_{1g} + B_{2g}$, $(e_u)^2 = A_{1g} + B_{1g} + B_{2g}$
$\mathbf{D_{5h}}$	$(e_1')^2 = A_1' + E_2'$, $(e_2')^2 = A_1' + E_1'$, $(e_1'')^2 = A_1' + E_2'$, $(e_2'')^2 = A_1' + E_1'$
$\mathbf{D_{6h}}$	$(e_{1g})^2 = A_{1g} + E_{2g}$, $(e_{2g})^2 = A_{1g} + E_{2g}$, $(e_{1u})^2 = A_{1g} + E_{2g}$, $(e_{2u})^2 = A_{1g} + E_{2g}$
$\mathbf{D_{\infty h}}$	$(\pi_g)^2 = \Sigma_g^+ + \Delta_g$, $(\pi_u)^2 = \Sigma_g^+ + \Delta_g$, $(\delta_g)^2 = \Sigma_g^+ + \Gamma_g$, $(\delta_u)^2 = \Sigma_g^+ + \Gamma_g$
$\mathbf{S_4}$	$(e)^2 = A + 2B$
$\mathbf{S_6}$	$(e_g)^2 = A_g + E_g$, $(e_u)^2 = A_g + E_g$
$\mathbf{S_8}$	$(e_1)^2 = A + E_2$, $(e_2)^2 = A + 2B$, $(e_3)^2 = A + E_2$
$\mathbf{T_d}$	$(e)^2 = A_1 + E$, $(t_1)^2 = A_1 + E + T_2$, $(t_2)^2 = A_1 + E + T_2$
$\mathbf{T}$	$(e)^2 = A + E$, $(t)^2 = A + E + T$
$\mathbf{O_h}$	$(e_g)^2 = A_{1g} + E_g$, $(e_u)^2 = A_{1g} + E_g$, $(t_{1g})^2 = A_{1g} + E_g + T_{2g}$, $(t_{2g})^2 = A_{1g} + E_g + T_{2g}$, $(t_{1u})^2 = A_{1g} + E_g + T_{2g}$, $(t_{2u})^2 = A_{1g} + E_g + T_{2g}$
$\mathbf{O}$	$(e)^2 = A_1 + E$, $(t_1)^2 = A_1 + E + T_2$, $(t_2)^2 = A_1 + E + T_2$

of inertia and one is the axis of minimum moment of inertia: these are necessarily perpendicular to each other and the third is then perpendicular to the other two. In cases where the principal axes are symmetry axes they are used as cartesian axes and labelled x, y, and z according to the conventions mentioned in sections 4.1 and 4.2(a). If three, two or none of the principal moments of inertia are equal the molecule is called respectively a *spherical, symmetric or asymmetric top.*

Symmetry species labels can be given to the processes of rotation R_x, R_y, or R_z of a molecule about the x-, y-, or z-axes respectively. Classification of the rotations of the water molecule is illustrated in figure 4.11 in which a + or – indicates whether the atom concerned is coming out of or going into the plane of the paper respectively. The rotations need only be classified according to their characters with respect to a set of generating elements, taken here to be C_2 and $\sigma_v(xz)$. It can be seen that $\Gamma(R_x) = B_2$, $\Gamma(R_y) = B_1$, and $\Gamma(R_z) = A_2$ which are indicated in the character table for the C_{2v} point group in table 4.11.

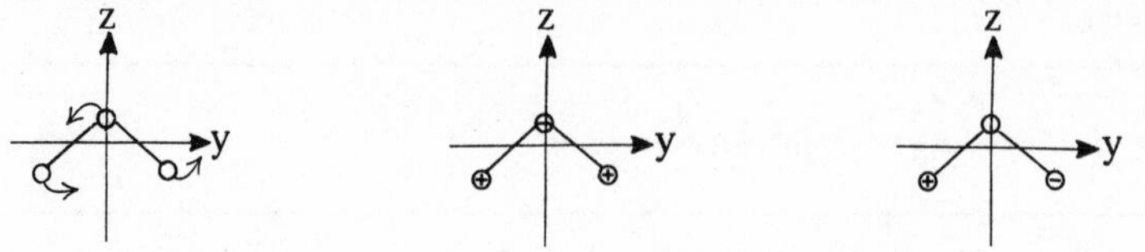

	C_2	$\sigma_v(xz)$
R_x	−1	−1

(a)

	C_2	$\sigma_v(xz)$
R_y	−1	1

(b)

	C_2	$\sigma_v(xz)$
R_z	1	−1

(c)

FIGURE 4.11
Symmetry classification of rotations about the principal axes in water

Assignments to symmetry species of translations T_x, T_y, and T_z of the water molecule along the cartesian axes are illustrated in figure 4.12. This figure demonstrates that $\Gamma(T_x) = B_1$, $\Gamma(T_y) = B_2$, and $\Gamma(T_z) = A_1$ which are also indicated in table 4.11.

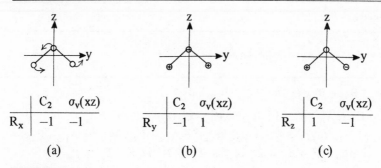

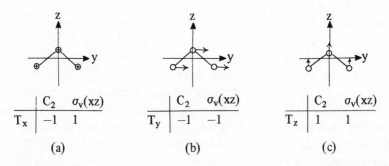

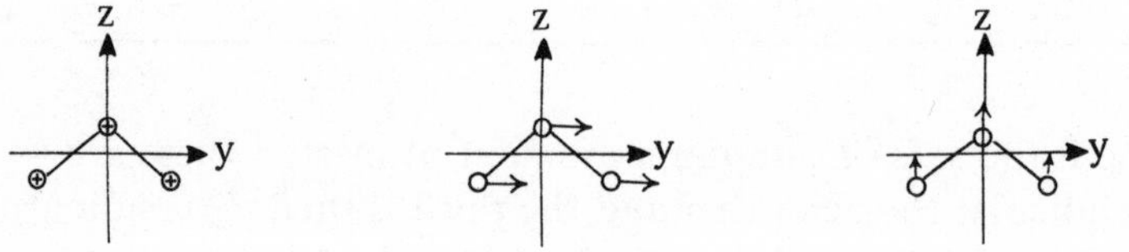

	C_2	$\sigma_v(xz)$
T_x	−1	1

(a)

	C_2	$\sigma_v(xz)$
T_y	−1	−1

(b)

	C_2	$\sigma_v(xz)$
T_z	1	1

(c)

FIGURE 4.12
Symmetry classification of translations along the principal axes in water

For all non-degenerate point groups the determination of the rotational and translational symmetry species is achieved by methods analogous to those used for the $\mathbf{C_{2v}}$ point group.

In degenerate point groups which do not contain triply degenerate symmetry species, that is all degenerate point groups which have been mentioned except $\mathbf{T_d}$, $\mathbf{T}$, $\mathbf{O_h}$, $\mathbf{O}$, and $\mathbf{K_h}$, there is only one unique axis, which is the z-axis. Rotation about and translation along this axis can be assigned to symmetry species in an analogous way to that used in non-degenerate point groups. For example figure 4.13(a) demonstrates that in methyl fluoride, which belongs to the $\mathbf{C_{3v}}$ point group, $\Gamma(R_z) = A_2$ and figure 4.13(b) shows that $\Gamma(T_z) = A_1$.

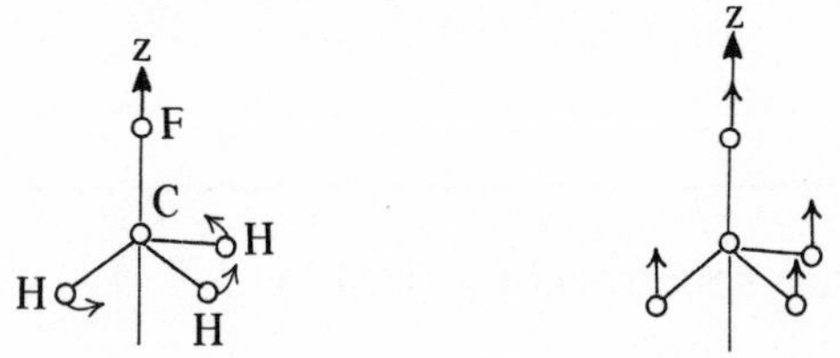

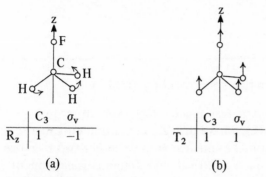

	C_3	σ_v
R_z	1	−1

(a)

	C_3	σ_v
T_z	1	1

(b)

FIGURE 4.13
Symmetry classification of (a) rotation about the z-axis and (b) translation along the z-axis in methyl fluoride

In any degenerate point group in which there is a unique z-axis, R_x and R_y collectively transform as one symmetry species, as do T_x and T_y, and the symmetry species must be a doubly degenerate one. In the $\mathbf{C_{3v}}$ point group there is only one degenerate symmetry species E and therefore the pairs (R_x, R_y) and (T_x, T_y) transform as E. In degenerate point groups with a unique axis but more than one degenerate symmetry species the species of (R_x, R_y) and (T_x, T_y) are not so obvious and will not be derived here. The results are given in the relevant character tables.

In the point groups $\mathbf{T_d}$, $\mathbf{T}$, $\mathbf{O_h}$, $\mathbf{O}$, and $\mathbf{K_h}$ there is no unique axis and R_x, R_y, and R_z must have the same species as must T_x, T_y, and T_z. This species is always a triply degenerate one. The assignments are given in the relevant character tables.

Assignments of components of the polarizability α to symmetry species will not be discussed until section 7.5 but these are also given in the character tables.

Some simple applications of molecular symmetry

5

5.1 Nuclear magnetic resonance spectroscopy (NMR)

As an analytical tool nuclear magnetic resonance (NMR) spectroscopy is able at best to distinguish nuclei which have different environments in a molecule. The nuclei must have non-zero nuclear spin. Whether two or more such nuclei, for example protons, have identical environments is dependent on their dispositions relative to the symmetry elements of the molecule.

We shall see that in practice one can say instinctively which nuclei in a molecule are in identical environments, without resorting to detailed symmetry arguments at all. However, it is useful to see why it is that instinct turns out to be correct, just as in the case of seeing why a circle is more symmetrical than a square. In addition, the application to NMR spectroscopy introduces a useful and important new concept—that of gross internal motions (other than vibration) taking place in a molecule during the time required to make an observation.

Although in chapter 1 it was stated that the total energy of a molecule is the sum of the rotational, vibrational and electronic energies, there may also be an additional energy due to nuclei having spin angular momentum. Nuclear spin angular momentum is quantized and has the value $[I(I+1)]^{\frac{1}{2}}\hbar$ where I is the nuclear spin quantum number which may be integral, half-integral or zero, $\hbar = h/2\pi$ and h is Planck's constant. If $I = 0$ there is no energy contributed by nuclear spin and there is no NMR spectrum produced by such nuclei. Some nuclei which have non-zero nuclear spin are listed in table 5.1. Of these nuclei the proton is of the greatest importance in NMR spectroscopy.

If $I \neq 0$ the nucleus has a magnetic moment μ and the angular momentum vector associated with it can take up $(2I+1)$ possible orientations in space

TABLE 5.1
Some nuclei with non-zero nuclear spin

Nucleus	*I*
^{1}H	$\frac{1}{2}$
^{2}H	1
^{13}C	$\frac{1}{2}$
^{14}N	1
^{19}F	$\frac{1}{2}$

such that the component along any particular direction, usually specified by the direction of a magnetic field, is given by $m_I\hbar$. m_I can take the values I, $I-1, I-2, \ldots -I$. We shall be considering here only the case of a proton for which $I = \frac{1}{2}$ and $m_I = +\frac{1}{2}$ or $-\frac{1}{2}$. The two states corresponding to the two values of m_I are of equal energy (degenerate) in the free molecule in the absence of a magnetic field. If a magnetic field B is introduced there is an energy $-\mu B m_I/[I(I+1)]^{\frac{1}{2}}$ associated with the nuclear magnetic moment

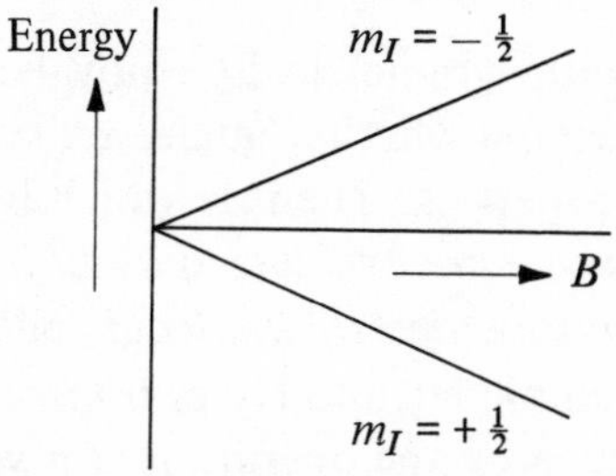

FIGURE 5.1
Variation of nuclear spin energy levels with magnetic field

and the two energy levels are split as shown in figure 5.1. Transitions between these two levels are allowed with a frequency given by

$$\nu = \mu B/[I(I+1)]^{1/2} h \qquad (5.1)$$

Thus a single proton in a magnetic field produces a single-line spectrum at a frequency given by equation 5.1. If there is more than one proton the intensity of the line is, to a first approximation, proportional to the number of protons and one might expect methyl alcohol ($CH_3 . OH$), for example, to give a single-line spectrum four times as intense as that of the single proton. In fact this is only very approximately true: the magnetic field experienced by a proton in a molecule is modified by its immediate environment of electrons. The effect of the environment is to modify the field experienced by the nucleus to $B(1-\delta)$. The frequency in equation 5.1 now becomes

$$\nu = \mu B(1-\delta)/[I(I+1)]^{1/2}\,h \tag{5.2}$$

and the shift in frequency from equation 5.1 of $-\mu B\delta/[I(I+1)]^{\frac{1}{2}}h$ is called the *chemical shift.* Protons in identical environments, that is protons which are equivalent by symmetry, experience the same chemical shift. An NMR spectrum of methyl alcohol ($CH_3.OH$) at low resolution is shown in figure 5.2(a). There are clearly two types of proton in this molecule and the integration of the area under each peak shows that there are three protons of one type and one of another type. Figure 5.2(b) shows a low resolution spectrum of ethyl alcohol ($CH_3.CH_2.OH$) which shows that there are three types of proton in the ratios 3:2:1. It should be noted that in figure 5.2 it is not frequency, but magnetic field which is plotted against intensity. It is experimentally more convenient to vary B and keep ν at a constant value in the range 50-100 MHz.

NMR spectroscopy, then, is concerned with symmetrically equivalent nuclei, especially protons. How do we determine whether nuclei are symmetrically equivalent by using symmetry properties? The rule which tells us the answer is *'If atoms (or nuclei) in a molecule are exchanged by at least one symmetry operation of the point group to which the molecule belongs, they are symmetrically equivalent'.* For example H_1 and H_2 in *trans*-difluoroethylene (figure 5.3(a)) are exchanged by the operation i (as well as by C_2). The three hydrogen atoms in ammonia (figure 5.3(b)) can all be exchanged by the σ_v operations. In naphthalene (figure 5.3(c)) $H_2 - H_3$ and $H_6 - H_7$ exchanges are effected by the $C_2(y)$ operation and $H_2 - H_7$ and $H_3 - H_6$ by $C_2(z)$. Therefore H_2, H_3, H_6, and H_7 are symmetrically equivalent as are H_1, H_4, H_5, and H_8. In 1-fluoro-3-chloro-5-bromobenzene (figure 5.3)d)) none of the three hydrogens can be exchanged by any symmetry operation of the group (C_s) and they are all non-equivalent.

From symmetry arguments applied so far it would not be expected that the three protons on the methyl group in methyl alcohol are equivalent.

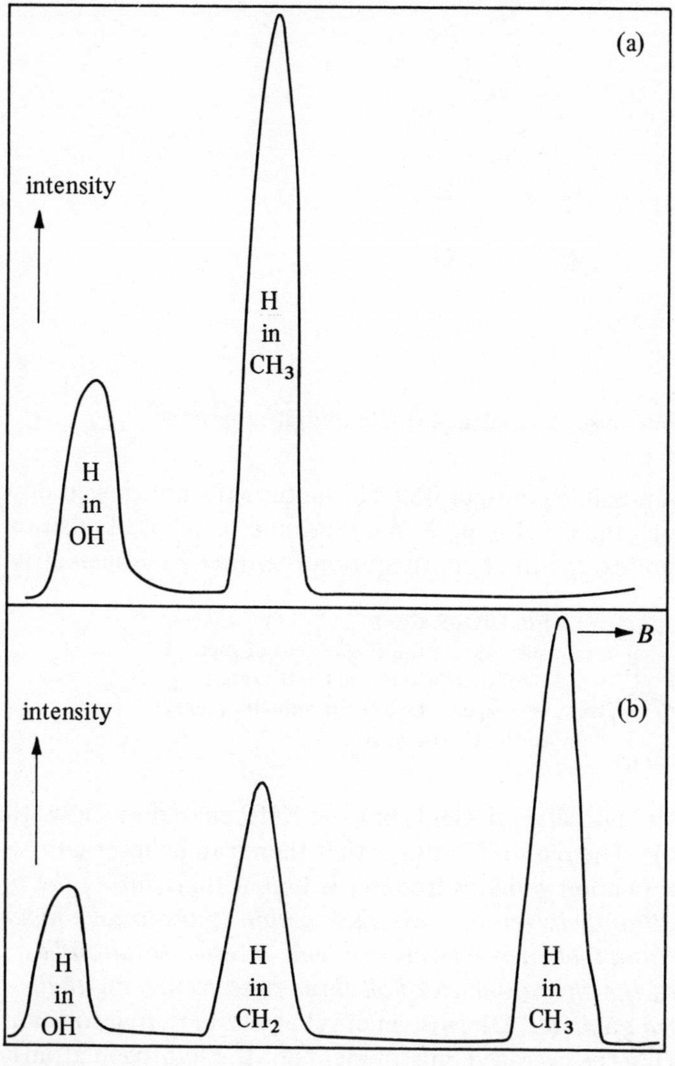

FIGURE 5.2
Low resolution NMR spectra of (a) methyl alcohol, (b) ethyl alcohol

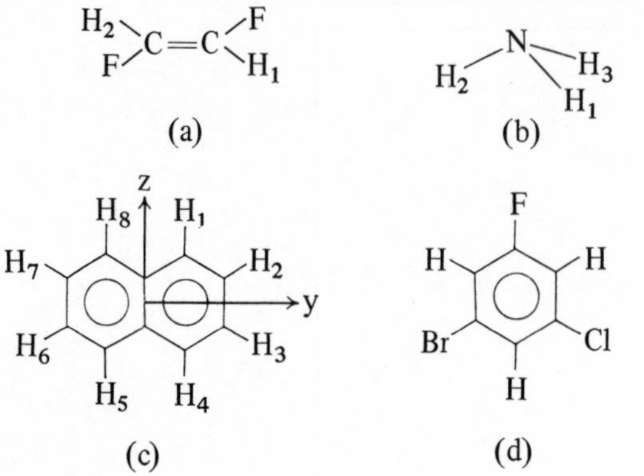

FIGURE 5.3
Illustration of some symmetrically equivalent and non-equivalent protons

Figure 5.4 shows two possible configurations of methyl alcohol viewed down the C—O bond in which the O—H bond is in a staggered or eclipsed position relative to the C—H bonds. In either configuration the three hydrogen atoms

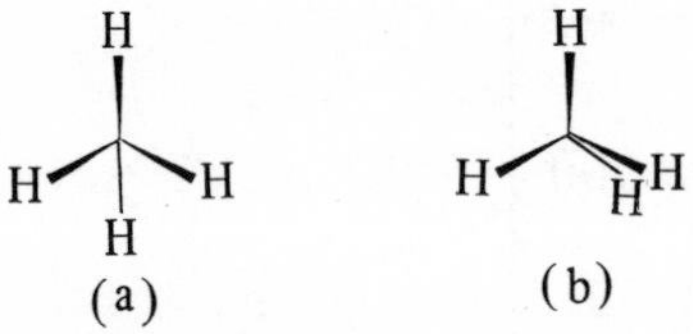

FIGURE 5.4
View down the C–O bond of methyl alcohol in which the O–H bond is (a) staggered, (b) eclipsed with respect to the C–H bonds

of the methyl group are not all equivalent, but the NMR spectrum shows that they are (figure 5.2(a)). The reason for this is that there can be internal rotation about the C—O bond which is free to the extent that, *during the time required for the transition between the two nuclear spin states to take place, transitions between rotational energy levels involving internal rotation has caused an averaging of the environments of all three protons making their effects in the spectrum identical.* Likewise in ethyl alcohol internal rotation has the effect of making the averaged environments of the hydrogen atoms of the CH_3 group identical and similarly for the hydrogen atoms of the CH_2 group.

In fact higher resolution spectra than those in figure 5.2 show that the spectrum is much more complex due to interactions between the nuclear spin angular momenta of the various protons namely spin-spin coupling. When higher resolution is used for methyl and ethyl alcohol, for example, each of the broad lines in figure 5.2 is split into several components, but the integrated intensity of the lines due to protons in a particular environment remains the same. If, however, spin-spin coupling causes splittings which are comparable with chemical shifts, the spectrum becomes still more complex. This occurs in molecules in which proton environments are similar, but not identical. For example, in fluorobenzene (figure 5.5) there are, from

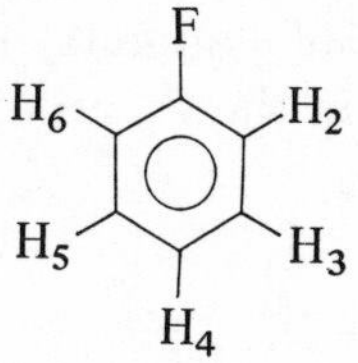

FIGURE 5.5
Fluorobenzene, in which all proton environments are quite similar

symmetry arguments, three types of proton, the pair $H_2 - H_6$, the pair $H_3 - H_5$, and H_4: but all the proton environments are sufficiently similar that spin-spin coupling is relatively large and a complex spectrum results from which it is not possible to see easily how many types of protons are present in the molecule.

5.2 Dipole moments

The components p_x, p_y, p_z of the electric dipole moment of a molecule along the cartesian axes are given by

$$p_x = \sum_i e_i x_i$$

$$p_y = \sum_i e_i y_i \tag{5.3}$$

$$p_z = \sum_i e_i z_i$$

where e_i is the charge and x_i, y_i, z_i the co-ordinates of the ith particle (nucleus or electron). If $p_x = p_y = p_z = 0$ the molecule has no permanent dipole moment. If only one of the components is non-zero it has a permanent dipole moment directed along one axis, which must be a symmetry axis. If two components are non-zero the permanent dipole moment is in a plane and all three components are non-zero if the dipole moment is not related to any of the principal axes.

The dipole moment is a vector and must be totally symmetric. The components p_x, p_y, p_z transform as T_x, T_y, T_z respectively and therefore, if any of the T_x, T_y, T_z are totally symmetric, the corresponding components of the dipole moment can be non-zero. These conditions require that molecules having a permanent dipole moment may belong only to the point groups $\mathbf{C}_1$, $\mathbf{C}_s$, $\mathbf{C}_n$, or $\mathbf{C}_{nv}$. In the $\mathbf{C}_n$ and $\mathbf{C}_{nv}$ point groups the dipole moment is along the C_n axis, in the $\mathbf{C}_s$ point group it lies in the plane of symmetry (in a direction which depends on the molecule concerned) and in the $\mathbf{C}_1$ point group it can lie in any direction (depending on the molecule).

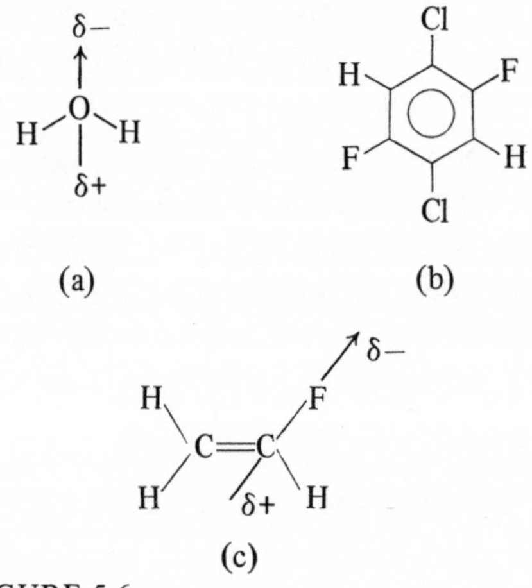

FIGURE 5.6
Some molecules with and without a dipole moment

Some examples are given in figure 5.6. The water molecule (figure 5.6(a)) belongs to the $\mathbf{C}_{2v}$ point group whose symmetry elements intersect in a line, the C_2 axis. Therefore this molecule has a permanent dipole moment along the C_2 axis. Symmetry arguments give no idea of the sense of the dipole, that is which end of the molecule has negative and which has positive charge. In practice, arguments concerning electronic structure often enable us to guess the direction of the dipole moment in what we believe to be a reliable manner. For example, in water it is expected that the sense of the dipole moment is that shown in figure 5.6(a); but, in a molecule like carbon monoxide (CO) for instance, which has a dipole moment of only 0.11D (= 0.37×10^{-30} Cm)† whereas HCl for example has a dipole moment of 1.08D (= 3.60×10^{-30} Cm), it is not possible to guess the sense of the dipole moment.

1,4-difluoro-2,5-dichlorobenzene (figure 5.6(b)) has a centre of symmetry and therefore no permanent dipole moment.

Monofluoroethylene (figure 5.6(c)) has only a plane of symmetry and it therefore has a permanent dipole moment which lies in this plane. The direction of the dipole is not defined by symmetry but probably lies near to the direction shown.

As in the case of determining equivalent hydrogen atoms in a molecule, one can usually say on inspection whether a molecule has a permanent dipole moment without using any detailed symmetry arguments, but it is useful to see how our conclusions can be rationalized by considerations of symmetry properties.

5.3 Optical activity

A molecule is said to be optically active if it rotates the plane of plane-polarized light passing through it.

In chemistry, optical activity is especially important in organic molecules. A criterion often used for ascertaining whether a molecule is optically active is to see if the molecule is superimposable on its mirror image. If it is, then the molecule is not optically active and *vice versa.* For example figure 5.7(a) shows that the molecule CHFClBr is not superimposable on its mirror image and is therefore optically active, but figure 5.7(b) shows that CH_2ClBr is not optically active.

A molecule like CHFClBr in which all four groups attached to the carbon atom are different is said to have an 'asymmetric carbon atom' and in simple organic molecules it is common to use the criterion of an asymmetric carbon atom to predict optical activity. However, in more complex molecules this criterion can be inadequate.

Consideration of symmetry properties leads to a very simple criterion which indicates whether a molecule is optically active. *If a molecule has an* S_n *axis of symmetry, where* $n = 1, 2, 3, \ldots,$ *it is not optically active while if*

† D is the symbol for debye; Cm is the symbol for coulomb metre.

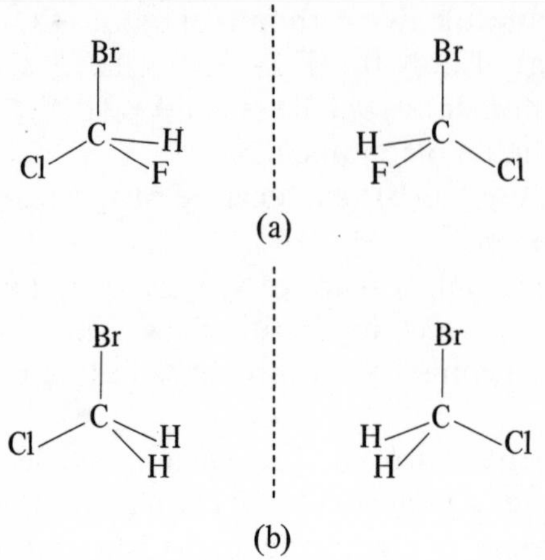

FIGURE 5.7
(a) An optically active and (b) an optically inactive substituted methane

it has no S_n *axis it is optically active.* Since $S_1 = \sigma$ and $S_2 = i$, any molecule having a plane or centre of symmetry is not optically active. The existence of a plane or centre of symmetry can be ascertained very easily and show a molecule (such as CH_2ClBr in figure 5.7(b)) to be optically inactive, but the existence of an S_n axis where $n > 2$ is not usually so obvious.

An example of the application of the criterion of the absence or presence of an S_n axis for deciding optical activity is the pair of molecules 1,1-difluoroallene and 1,3-difluoroallene illustrated in figures 5.8(a) and (b) respectively.

FIGURE 5.8
(a) 1,1-Difluoroallene, which is optically inactive, and (b) 1,3-difluoroallene, which is optically active

1,1-Difluoroallene has a plane of symmetry (it belongs to the $\mathbf{C}_{2v}$ point group) and so is not optically active, while 1,3-difluoroallene has only a C_2 axis of symmetry (it belongs to the $\mathbf{C}_2$ point group) and so is optically active.

An example of a molecule which is not optically active because it has an S_4 axis but no i or σ symmetry element is the tetramethyl spiropentane illustrated in figure 5.9.

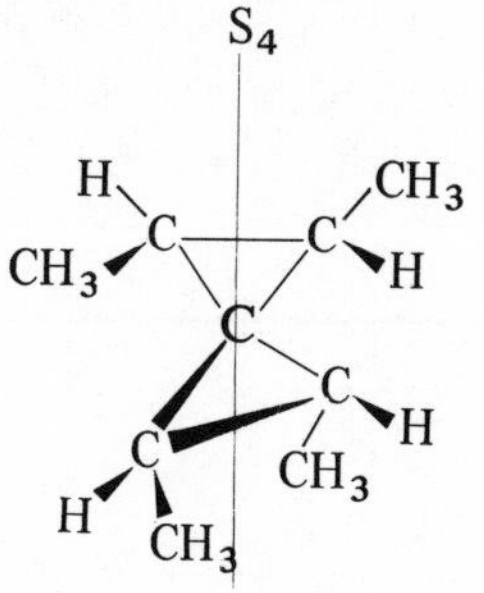

FIGURE 5.9
Tetramethyl spiropentane has an S_4 axis and is therefore optically inactive in spite of not having a plane or centre of symmetry

5.4 Effects of isotopic substitution

In chapter 3 the assignment of molecules to point groups was discussed in detail. These are the point groups to which the equilibrium nuclear configurations belong. Thus benzene is assigned to the $\mathbf{D_{6h}}$ point group and mono-deuterobenzene to the $\mathbf{C_{2v}}$ point group. However, if we are concerned with a property such as the electronic wave function all isotopic species can be taken as equivalent: the form of the electronic wave function is dependent on nuclear charges and the number of electrons in a molecule, and not on nuclear masses. The electronic wave functions of mono-deuterobenzene should be classified, therefore, according to the $\mathbf{D_{6h}}$ point group and not $\mathbf{C_{2v}}$. On the other hand vibrational wave functions are highly sensitive to nuclear masses and so must be classified according to the $\mathbf{C_{2v}}$ point group.

Similarly the electronic and vibrational wave functions of *trans*-1,2-di-deuteroethylene should be classified according to the $\mathbf{D_{2h}}$ and $\mathbf{C_{2h}}$ point groups respectively.

Introduction to molecular orbital methods 6

6.1 LCAO method

Before applying molecular symmetry arguments to electronic wave functions it will be necessary to have some qualitative idea of the nature of these wave functions.

The method used most commonly for determining approximate electronic wave functions is one of the molecular orbital (MO) methods and in particular the linear combination of atomic orbitals (LCAO) MO method.

MO's have properties which are similar to those of atomic orbitals (AO's) in that (a) an MO can be described by a wave function ψ which may cover the whole of the molecule; (b) if ψ is normalized i.e. $\int \psi^* \psi d\tau = 1$†, then $\psi^* \psi d\tau$ represents the probability of finding the electron in the element of volume $d\tau$; (c) the wave function may have quantum numbers associated with it and the shape and size of the orbital depends on their values; (d) there is a spin quantum number m_s $(= \pm \frac{1}{2})$ associated with each electron in an MO; (e) in an analogous way to atoms, the electrons in molecules can be fed into the calculated MO's in pairs, one with $m_s = +\frac{1}{2}$ and one with $m_s = -\frac{1}{2}$, in order of increasing energy to give the ground state configuration of the molecule.

The LCAO method of constructing MO wave functions takes account of the fact that for the region of the MO in the neighbourhood of a nucleus, the wave function will resemble an AO wave function of the atom of which the

† ψ^* is the complex conjugate of ψ and is obtained from it by replacing all $i(=\sqrt{-1})$ by $-i$.

nucleus is a part. It seems reasonable, therefore, to express the MO wave function ψ as a linear combination of AO wave functions χ_i

$$\begin{aligned}\psi &= c_1\chi_1 + c_2\chi_2 + \ldots c_n\chi_n \\ &= \sum_i c_i\chi_i\end{aligned} \tag{6.1}$$

where the c_i are coefficients of the AO wave functions. However, not all linear combinations are effective, in that some would result in MO's which are very little different in energy from the AO's from which they are formed: this would result in one of the coefficients c_i being much greater than all the others. For linear combinations to be effective three conditions must be satisifed

(*i*) The energies of the AO's must be comparable.
(*ii*) The AO's should overlap as much as possible.
(*iii*) The AO's must have the same symmetry properties with respect to certain symmetry elements of the molecule.

We will now follow through the evaluation of the energies and wave functior of two MO's constructed from two AO's in the case of a diatomic molecule.

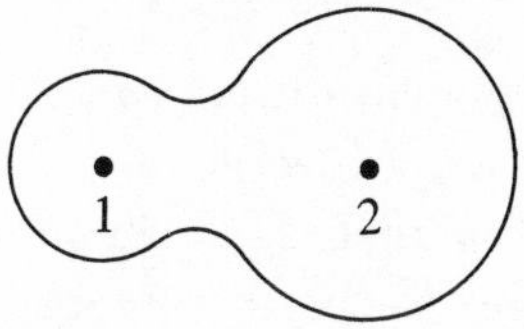

FIGURE 6.1
A molecular orbital around two nuclei 1 and 2

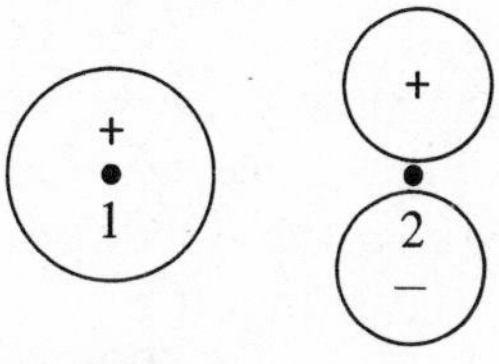

FIGURE 6.2
An s orbital on 1 cannot be linearly combined with this p orbital on 2 to form a molecular orbital

The two nuclei (figure 6.1) are labelled 1 and 2 and we expect the MO in the region of nucleus 1 to resemble an AO of atom 1 and in the region of nucleus 2 to resemble an AO of atom 2. The LCAO method then gives the MO wave function

$$\psi = c_1\chi_1 + c_2\chi_2 \tag{6.2}$$

provided that χ_1 and χ_2 satisfy the three conditions above. As an example of condition (*iii*) the s and p orbitals in figure 6.2 cannot form an effective linear combination since the s orbital is symmetric, and the p orbital antisymmetric with respect to reflection in the σ_v plane perpendicular to the figure and containing the internuclear axis.

In order to determine c_1, c_2 and E, the energy associated with the MO, we need to use the *variation method*. This method enables us to obtain the best values of c_1 and c_2 for the MO.

The MO wave function must satisfy the wave equation

$$\mathrm{H}\psi = \mathrm{E}\psi \tag{6.3}$$

where H is the Hamiltonian operator which operates on ψ to give the product of ψ and the total energy E associated with the MO. If we multiply both sides by ψ^* and integrate we get

$$E = \int \psi^* \mathrm{H}\psi \, \mathrm{d}\tau \Big/ \int \psi^* \psi \, \mathrm{d}\tau \tag{6.4}$$

So, if we know ψ and H, we can calculate E but, if ψ is not known, we can make an intelligent guess at it, say ψ_n, and calculate a quantity $\bar{E}_n$, the expectation value of the energy

$$\bar{E}_n = \int \psi_n^* \mathrm{H}\psi_n \, \mathrm{d}\tau \Big/ \int \psi_n^* \psi_n \, \mathrm{d}\tau \tag{6.5}$$

$\bar{E}_n$ is an expectation value because, since ψ_n is not the true wave function, we are not calculating the energy of a stationary state of the molecule.

We can guess again at the wave function, say ψ_m, and calculate $\bar{E}_m$. The variation principle states that if $\bar{E}_m < \bar{E}_n$ then $\bar{E}_m$ is closer to the true energy and ψ_m closer to the true wave function than are $\bar{E}_n$ and ψ_n. In practice the trial wave functions are not guessed at random, but are expressed in terms of variable parameters (c_1 and c_2 in the present case) whose optimum values are found using the variation method.

From equations 6.2 and 6.5 we have

$$\begin{aligned}\bar{E} &= \frac{\int (c_1\chi_1^* + c_2\chi_2^*)\,\mathrm{H}(c_1\chi_1 + c_2\chi_2)\,\mathrm{d}\tau}{\int (c_1\chi_1^* + c_2\chi_2^*)(c_1\chi_1 + c_2\chi_2)\,\mathrm{d}\tau} \\ &= \frac{\int (c_1^2\chi_1^*\mathrm{H}\chi_1 + c_1c_2\chi_1^*\mathrm{H}\chi_2 + c_1c_2\chi_2^*\mathrm{H}\chi_1 + c_2^2\chi_2^*\mathrm{H}\chi_2)\,\mathrm{d}\tau}{\int (c_1^2\chi_1^*\chi_1 + c_1c_2\chi_1^*\chi_2 + c_1c_2\chi_1\chi_2^* + c_2^2\chi_2^*\chi_2)\,\mathrm{d}\tau}\end{aligned} \tag{6.6}$$

Now, if the AO wave functions are normalized, $\int \chi_1^* \chi_1 d\tau = \int \chi_2^* \chi_2 d\tau = 1$ and if H is a Hermitian operator, which will always be the case, $\int \chi_1^* H \chi_2 d\tau = \int \chi_2^* H \chi_1 d\tau = H_{12}$, say. If we also make the substitutions $\int \chi_1^* H \chi_1 d\tau = H_{11}$, $\int \chi_2^* H \chi_2 d\tau = H_{22}$ and $\int \chi_1^* \chi_2 d\tau = \int \chi_1 \chi_2^* d\tau = S_{12}$, the overlap integral, which is a measure of the overlap of χ_1 and χ_2, equation 6.6 becomes

$$\bar{E} = \frac{c_1^2 H_{11} + 2c_1 c_2 H_{12} + c_2^2 H_{22}}{c_1^2 + 2c_1 c_2 S_{12} + c_2^2} \tag{6.7}$$

Obtaining $\partial \bar{E}/\partial c_1$ from equation 6.7 and putting it equal to zero gives, from the variation principle, the value of c_1 which corresponds to the lowest value of $\bar{E}$ (which we now call E) from the expression

$$c_1(H_{11} - E) + c_2(H_{12} - ES_{12}) = 0 \tag{6.8}$$

Similarly, obtaining $\partial \bar{E}/\partial c_2$ from equation 6.7 gives

$$c_1(H_{12} - ES_{12}) + c_2(H_{22} - E) = 0 \tag{6.9}$$

Equations 6.8 and 6.9 are called the *secular equations.* The two values of E which satisfy these equations are obtained from the *secular determinant*

$$\begin{vmatrix} H_{11} - E & H_{12} - ES_{12} \\ H_{12} - ES_{12} & H_{22} - E \end{vmatrix} = 0 \tag{6.10}$$

If we consider the case of a homonuclear diatomic molecule (where the two nuclei are identical) then $H_{11} = H_{22}$ which we call α, the Coulomb integral. H_{12}, the resonance integral, is often labelled β. Then we have

$$\begin{vmatrix} \alpha - E & \beta - ES \\ \beta - ES & \alpha - E \end{vmatrix} = 0 \tag{6.11}$$

Solving for E gives the two values

$$E_{\pm} = \frac{\alpha \pm \beta}{1 \pm S} \tag{6.12}$$

If we make two further approximations (a) that $S = 0$ (in fact a typical value is about 0.2), and (b) that H is taken to be the same as in the atom, then $\alpha = E_A$, the AO energy, and

$$E_{\pm} = E_A \pm \beta \tag{6.13}$$

The energies $E_{\pm}$ which result from the LCAO treatment, with the further approximations, are then symmetrically disposed about the AO energy E_A. The process of the energies of the two MO's being pushed apart, relative to those of the AO's, is a resonance process and it is for this reason that β is called the *resonance integral.* As is clear from figure 6.3, β is a negative quantity.

It is a general rule that a linear combination of two AO's will give two MO's, one higher and the other lower in energy than either of the two AO's. The closer together are the energies of the two AO's the greater will be the 'resonance' between them and the larger β will be. Conversely, if β is close

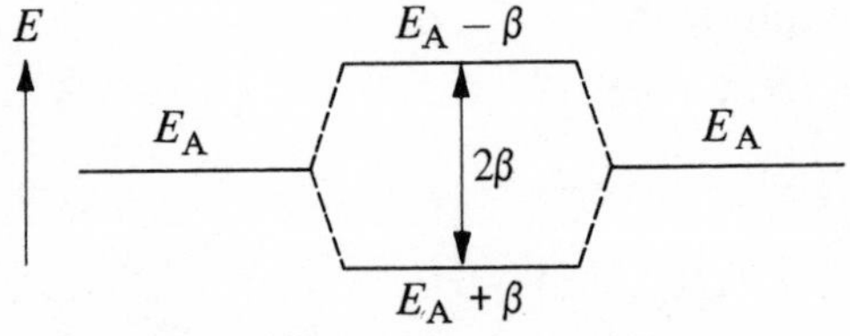

FIGURE 6.3
Energy level diagram for two MO's formed from two identical AO's

to zero, the two AO energies must be very different and the linear combination is not an effective one: hence the rule that in an LCAO treatment the AO's must be of comparable energies.

With the approximation that $S = 0$, the secular determinant of equation 6.11 becomes

$$\begin{vmatrix} \alpha - E & \beta \\ \beta & \alpha - E \end{vmatrix} = 0 \tag{6.14}$$

and the corresponding secular equations are

$$\begin{aligned} c_1(\alpha - E) + c_2\beta &= 0 \\ c_1\beta + c_2(\alpha - E) &= 0 \end{aligned} \tag{6.15}$$

Putting $E = E_+$ or E_- we get $c_1/c_2 = 1$ or -1 respectively and therefore the wave functions ψ_+ and ψ_- corresponding to E_+ and E_- are given by

$$\begin{aligned} \psi_+ &= N_+(\chi_1 + \chi_2) \\ \psi_- &= N_-(\chi_1 - \chi_2) \end{aligned} \tag{6.16}$$

where N_+ and N_- are normalization constants obtained from the conditions $\int \psi_+^2 d\tau = \int \psi_-^2 d\tau = 1$. Neglecting $\int \chi_1 \chi_2 d\tau$, the overlap integral, we get $N_+ = N_- = 1/\sqrt{2}$ and

$$\psi_\pm = (\chi_1 \pm \chi_2)/\sqrt{2} \tag{6.17}$$

If χ_1 and χ_2 are 1s wave functions then the ψ_+ wave function has no node between the two nuclei, while ψ_- has one node. This is illustrated in figure 6.4

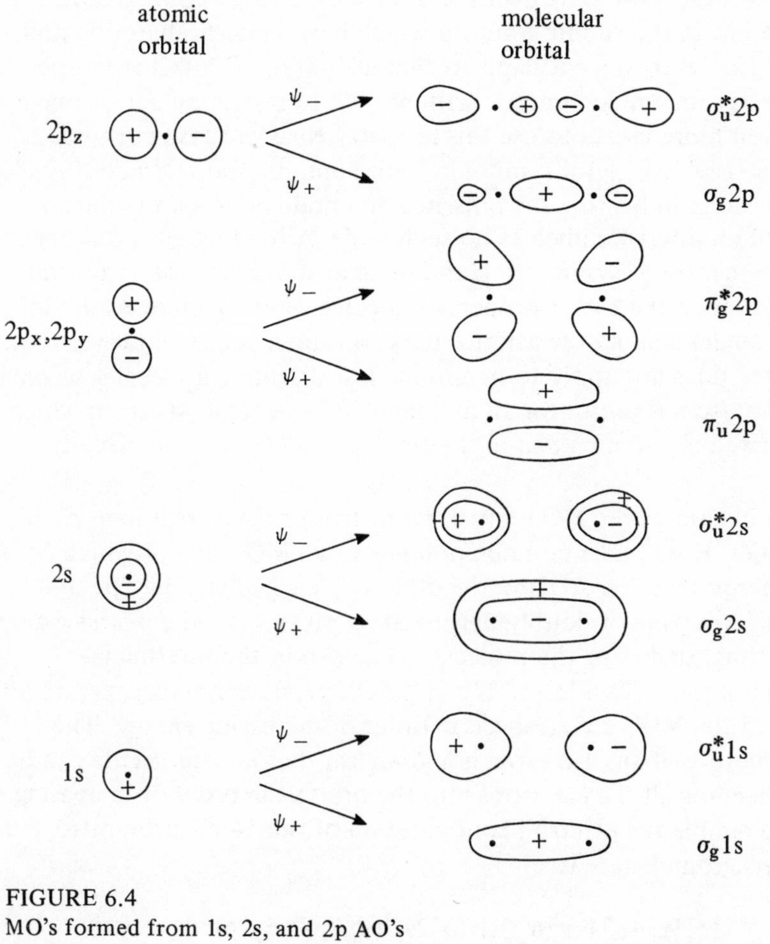

FIGURE 6.4
MO's formed from 1s, 2s, and 2p AO's

for the $\sigma_g 1s$ and $\sigma_u^* 1s$ orbitals resulting from an LCAO treatment of 1s AO's. In this case the MO wave functions are given by

$$\psi_\pm = [\chi_1(1s) \pm \chi_2(1s)]/\sqrt{2} \tag{6.18}$$

In the labelling of MO's, σ and π indicate that the quantum number λ associated with the component of the electronic orbital angular momentum along the internuclear z-axis has the values 0 and 1 respectively. This convention was mentioned in section 4.2(c) with reference to the $\mathbf{C}_{\infty v}$ point group, but it is used also in the $\mathbf{D}_{\infty h}$ point group to which homonuclear diatomic molecules belong. However, since it happens that all σ and π orbitals are respectively symmetric and antisymmetric with respect to reflection across one σ_v plane it is often more useful to use this property for identifying orbitals as σ or π. The asterisk is a useful symbol in both homonuclear and heteronuclear diatomic molecules indicating the presence of a node between the nuclei: the absence of an asterisk indicates no such node. A homonuclear diatomic molecule has a centre of symmetry i, and the g or u property of an orbital in combination with the σ or π property indicates whether there is a node between the nuclei making the asterisk notation superfluous. However the g or u property does not apply to hetero-nuclear diatomic molecules whereas the asterisk notation is useful for all diatomics. The type of AO from which the MO is derived is also included in the label, e.g. $\sigma_u^* 1s$ or, more usually, $\sigma^* 1s$.

Figure 6.4 illustrates the MO's which result from LCAO treatment of 1s, 2s, and 2p AO's. Each linear combination gives two MO's one of which (ψ_+) is lower in energy than the AO and the other (ψ_-) is higher. The $2p_x$ and $2p_y$ AO's give two pairs of doubly degenerate π MO's since the x- and y-axes are still indistinguishable in the molecule. The z-axis, the internuclear axis, is unique in the molecule and the $2p_z$ AO's produce non-degenerate MO's.

In figure 6.5 the MO's are arranged in order of increasing energy. The electronic structure of any first-row homonuclear diatomic molecule can be obtained by feeding all the electrons into the orbitals in order of increasing energy. For example the electron configuration of the 14-electron nitrogen molecule in its ground state is

$$(\sigma_g 1s)^2(\sigma_u^* 1s)^2(\sigma_g 2s)^2(\sigma_u^* 2s)^2(\pi_u 2p)^4(\sigma_g 2p)^2$$

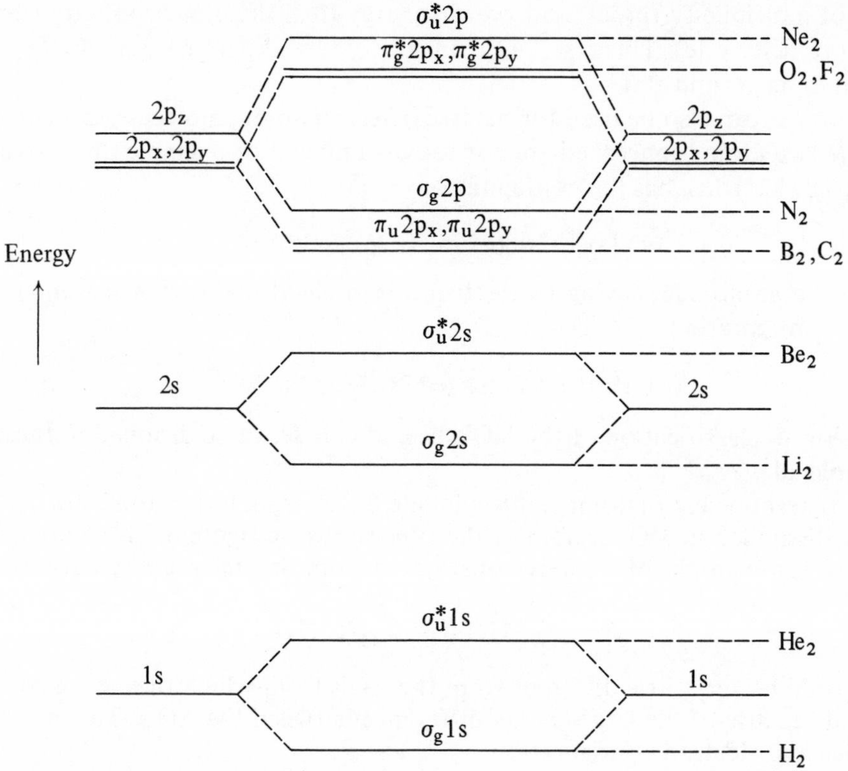

FIGURE 6.5
MO energy level diagram for first-row homonuclear diatomic molecules

and of the 16-electron oxygen molecule is

$$(\sigma_g 1s)^2(\sigma_u^* 1s)^2(\sigma_g 2s)^2(\sigma_u^* 2s)^2(\pi_u 2p)^4(\sigma_g 2p)^2(\pi_g^* 2p)^2$$

In oxygen the two electrons in the π_g^*2p orbital can go either, with spins antiparallel, into one of the degenerate orbitals or, with spins parallel, one into each of the degenerate orbitals: neither situation violates the Pauli principle. However, one of Hund's rules says that, if a number of configurations are equivalent except in respect of their multiplicity, the configuration of highest multiplicity lies lowest in energy. In the case of oxygen, therefore, the configuration in which the two electrons in the π_g^*2p orbital have parallel spins (of multiplicity three) lies lower in energy than the other configuration (of multiplicity one). This result explains the observed paramagnetism of oxygen in its ground state.

Figure 6.5 can also be used for heteronuclear diatomic molecules provided that the two atoms concerned are not too dissimilar. For example nitric oxide, having 15 electrons, has the configuration

$$(\sigma 1s)^2(\sigma^* 1s)^2(\sigma 2s)^2(\sigma^* 2s)^2(\pi 2p)^4(\sigma 2p)^2(\pi^* 2p)$$

and carbon monoxide, having 14 electrons, is isoelectronic with nitrogen and has the configuration

$$(\sigma 1s)^2(\sigma^* 1s)^2(\sigma 2s)^2(\sigma^* 2s)^2(\pi 2p)^4(\sigma 2p)^2$$

The 'g' or 'u' classification of the MO's has, of course, to be dropped in these two molecules.

In a heteronuclear diatomic molecule like HCl in which the atoms are grossly dissimilar an MO diagram of the type shown in figure 6.5 is of no use at all. If, for example, MO's were constructed from the linear combination

$$\psi = c_1 \chi_1(\mathrm{H{:}1s}) + c_2 \chi_2(\mathrm{Cl{:}1s})$$

they would be very little different from the AO's from which they were constructed because of the large energy difference between the AO's. The chlorine atom has the electron configuration

$$1s^2\,2s^2\,2p^6\,3s^2\,3p^5$$

and the useful linear combination to make is between the hydrogen 1s and the chlorine $3p_z$ AO's: these are of similar energies and have the same symmetry with respect to one of the σ_v planes. Hence the MO's into which the two electrons in the bond can be fed are given by

$$\psi = c_1 \chi_1(\mathrm{H{:}1s}) + c_2 \chi_2(\mathrm{Cl{:}3p_z})$$

This linear combination gives a σ and a σ^* MO and the two electrons go, in the ground state of HCl, into the σ MO. Four electrons on the chlorine atom remain in the $2p_x$ and $2p_y$ orbitals, which are virtually unchanged from those in the atom: electrons in these orbitals are called *lone pair* electrons.

In a general case of the LCAO method in a polyatomic molecule the MO wave function is given by equation 6.1 and the secular determinant analogous to equation 6.10 becomes

$$\begin{vmatrix} H_{11}-E & H_{12}-ES_{12} \cdots & H_{1n}-ES_{1n} \\ H_{12}-ES_{12} & H_{22}-E \quad \cdots & H_{2n}-ES_{2n} \\ \vdots & \vdots & \vdots \\ H_{1n}-ES_{1n} & H_{2n}-ES_{2n} \cdots & H_{nn}-E \end{vmatrix} = 0 \qquad (6.19)$$

which can be written

$$\left| H_{\mu\nu} - ES_{\mu\nu} \right| = 0 \qquad (6.20)$$

From this determinant the coefficients c_i in equation 6.1 can be determined provided that the forms of the $H_{\mu\nu}$ are known which requires that the form of the Hamiltonian operator which is involved in these integrals is known.

The true Hamiltonian $\mathrm{H_e}$ for a many-electron system can be expressed as

$$\mathrm{H_e} = \sum_{\mu} \mathrm{H}_{\mu} + \sum_{\mu>\nu}\sum e^2/r_{\mu\nu} \qquad (6.21)$$

where H_{μ} is the part of a one-electron Hamiltonian which does not include electron repulsions: these are all included in the second term.

The concept of MO's into which we can feed electrons, irrespective of their number, in order of increasing energy until the electrons are used up, is dependent on some method of approximating the electron repulsion term in $\mathrm{H_e}$ to a sum of contributions by each electron. One approximation involves regarding the electron repulsions as an averaged potential field, $U(r_{\mu})$, in which the electrons move: then we have

$$\mathrm{H_e} = \sum_{\mu} \mathrm{H}_{\mu} + \sum_{\mu} U(r_{\mu}) = \sum_{\mu} \mathcal{H}_{\mu} \qquad (6.22)$$

where r_{μ} is a co-ordinate of the μth electron. The eigenfunctions of the one-electron approximate Hamiltonians $\mathcal{H}_{\mu}$ are the one-electron MO wave functions $\phi_i(r_{\mu})$ which are required and are determined from

$$\mathcal{H}_{\mu}\,\phi_i(r_{\mu}) = E_i\,\phi_i(r_{\mu}) \qquad (6.23)$$

and the total energy is given by

$$E = \sum_{i} E_i \qquad (6.24)$$

for all the filled orbitals.

The method of approximating the electron repulsions by a sum of one-electron contributions to an averaged potential field is the basis of the Hartree-Fock self-consistent field (SCF) MO method.

6.2 Hückel molecular orbitals

The Hartree-Fock method outlined above is unfortunately very difficult to apply to molecules with a large number of electrons. However, in the case of molecules with conjugated π-electron systems, for example butadiene and benzene, a very approximate method of obtaining the higher energy MO's, the Hückel MO method, has had considerable success in describing these MO's in a semi-quantitative way. The method has been modified and extended but there were five approximations involved in the original Hückel method

(*i*) π-Electrons only are considered, since it is assumed that π MO's are very much higher in energy than σ MO's and therefore can be treated separately.

(*ii*) Zero overlap of AO's is assumed even between neighbouring atoms giving, from equation 6.20

$$S_{\mu\nu} = 0 \tag{6.25}$$

unless $\mu = \nu$ in which case $S_{\mu\nu} = 1$.

(*iii*) $H_{\mu\mu}$ is assumed to be the same for each atom (usually carbon, nitrogen or oxygen) irrespective of its environment giving

$$H_{\mu\mu} = \alpha \tag{6.26}$$

say, where α is called the Coulomb integral as in section 6.1.

(*iv*) The resonance integral $H_{\mu\nu}$ is assumed to be the same for any pair of atoms μ and ν directly bonded to each other. As in section 6.1, the substitution

$$H_{\mu\nu} = \beta \tag{6.27}$$

is usually made.

(*v*) $H_{\mu\nu}$ is assumed to be zero if μ and ν are not directly bonded to each other.

π-Electron MO wave functions in the Hückel method are given as in equation 6.1 by

$$\psi = \sum_i c_i \chi_i \tag{6.28}$$

but now the χ_i are wave functions of only those AO's, usually 2p, which form the π MO's. The secular determinant of equation 6.19 is used to determine the MO energies.

In the simple case of ethylene there are, after formation of the C—C and C—H σ-bonds, two electrons, one from each carbon atom, in 2p orbitals perpendicular to the plane of the molecule, as shown in figure 6.6, which

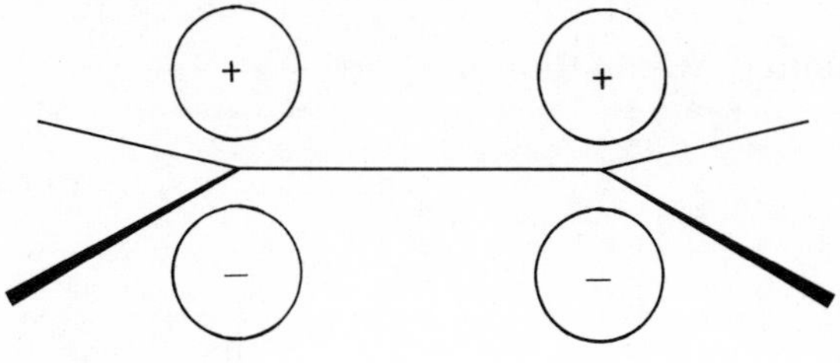

FIGURE 6.6
2p Atomic orbitals on the carbon atoms of ethylene directed perpendicular to the molecular plane

are available for π-bond formation. In the formation of MO's from the two carbon 2p AO's the MO wave function is

$$\psi = c_1 \chi_1 + c_2 \chi_2 \tag{6.29}$$

and equation 6.19 becomes, with the assumptions and substitutions mentioned in (*i*)-(*v*),

$$\begin{vmatrix} \alpha - E & \beta \\ \beta & \alpha - E \end{vmatrix} = 0 \tag{6.30}$$

Dividing through by β and putting $(\alpha - E)/\beta = x$ we get

$$\begin{vmatrix} x & 1 \\ 1 & x \end{vmatrix} = 0 \tag{6.31}$$

from which it follows that $x = \pm 1$ and $E = \alpha \pm \beta$, with the result that there are two πMO's. β is a negative quantity and is of the order of $-70\,000\ \text{cm}^{-1}$ ($= -84\ \text{kJ mol}^{-1}$) in carbon systems. Unless the total energy is required, α can be taken as the zero of energy and the two MO's are then symmetrically disposed about zero as shown in figure 6.7.

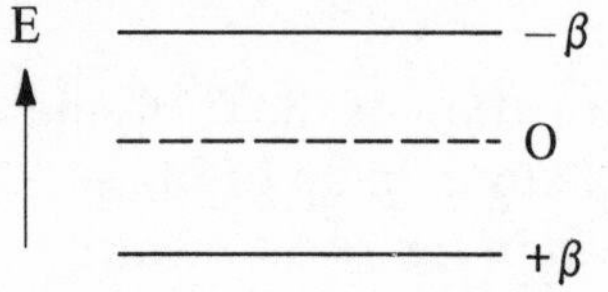

FIGURE 6.7
π-MO energy level diagram for ethylene

The secular equations represented by equation 6.31 are

$$\begin{aligned} c_1 x + c_2 &= 0 \\ c_1 + x c_2 &= 0 \end{aligned} \qquad (6.32)$$

Taking the value $x = -1$

$$c_1 - c_2 = 0 \qquad (6.33)$$

and from the normalization of the wave function ($\int \psi^* \psi \, d\tau = 1$) we have

$$c_1^2 + c_2^2 = 1 \qquad (6.34)$$

From equations 6.33 and 6.34 it follows that $c_1 = c_2 = 1/\sqrt{2}$ and the wave function for $x = -1$ and $E = \alpha + \beta$ is

$$\psi_1 = (\chi_1 + \chi_2)/\sqrt{2} \qquad (6.35)$$

Similarly for $x = 1$ and $E = \alpha - \beta$, the wave function is

$$\psi_2 = (\chi_1 - \chi_2)/\sqrt{2} \qquad (6.36)$$

The wave function ψ_2 has a node in between the two carbon atoms whereas ψ_1 has no node. The forms of these wave functions are illustrated in figure 6.8.

It is useful at this stage to classify the wave functions according to symmetry species of the point group $\mathbf{D}_{2h}$, to which ethylene belongs. Using the character table given in table 4.32 and the conventional axis notation in which the C—C axis is the z-axis, and the x-axis is perpendicular to the molecular plane, it is easily shown that ψ_1 and ψ_2 have symmetry species b_{3u} and b_{2g} respectively. (It should be noted here that it is conventional for lower case letters to be used to denote orbital symmetry species). The configuration of the two π-electrons in the ground state of ethylene is therefore

$$(b_{3u})^2$$

It is a useful shorthand convention to denote ψ_1 by π (antisymmetric to the plane of the moelcule) and ψ_2 by π^*, where the asterisk indicates the antibonding character of ψ_2 due to the node between the carbon atoms. Thus

FIGURE 6.8
Two π-orbitals of ethylene

the electron promotion in going from the ground to the first excited state is often given as $\pi^* - \pi$ rather than $b_{2g} - b_{3u}$. (It is conventional in electronic transitions to put the orbital of higher energy first and that of lower energy second, but the opposite is often used. The use of an arrow instead of a dash is sometimes convenient to indicate an absorption process e.g. $\pi^* \leftarrow \pi$ or an emission process e.g. $\pi^* \rightarrow \pi$).

Butadiene has a conjugated π-electron system and can also be treated by the Hückel method. This molecule has the *s-trans* structure (*trans* about the C—C single bond) shown in figure 6.9 but, in the Hückel method, we can regard the carbon chain as linear since the method treats all the carbon atoms as equivalent.

π MO's in butadiene are formed from four 2p AO's, one on each carbon atom and directed perpendicular to the molecular plane as shown in figure 6.10. The MO wave functions are given by

$$\psi = c_1\chi_1 + c_2\chi_2 + c_3\chi_3 + c_4\chi_4 \qquad (6.37)$$

The resulting secular determinant, analogous to that of equation 6.31 for ethylene, is

$$\begin{vmatrix} x & 1 & 0 & 0 \\ 1 & x & 1 & 0 \\ 0 & 1 & x & 1 \\ 0 & 0 & 1 & x \end{vmatrix} = 0 \tag{6.38}$$

Solving this determinant gives

$$x = \pm 1.62 \text{ or } \pm 0.62$$

The energy-level diagram for the four resulting MO's is given in figure 6.11. The coefficients c_i in equation 6.37 can be obtained from the secular equations

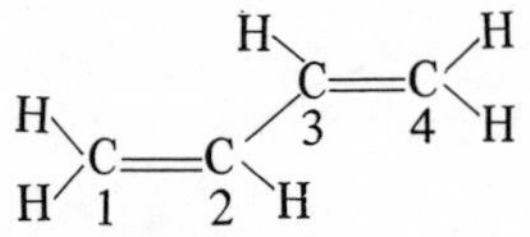

FIGURE 6.9
s-trans Butadiene

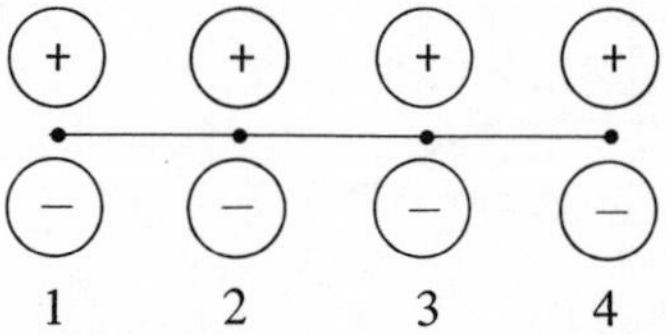

FIGURE 6.10
2p Atomic orbitals on the four carbon atoms directed perpendicular to the molecular plane

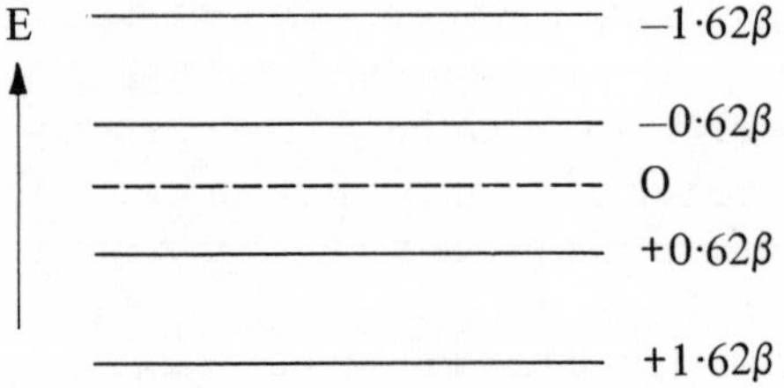

FIGURE 6.11
π-MO energy level diagram for butadiene

corresponding to the determinant of equation 6.38 and the normalization condition for ψ. The four MO wave functions are then shown to be

$$\begin{aligned} \psi_1 &= 0.37\chi_1 + 0.60\chi_2 + 0.60\chi_3 + 0.37\chi_4 \\ \psi_2 &= 0.60\chi_1 + 0.37\chi_2 - 0.37\chi_3 - 0.60\chi_4 \\ \psi_3 &= 0.60\chi_1 - 0.37\chi_2 - 0.37\chi_3 + 0.60\chi_4 \\ \psi_4 &= 0.37\chi_1 - 0.60\chi_2 + 0.60\chi_3 - 0.37\chi_4 \end{aligned} \tag{6.39}$$

The forms of these wave functions are illustrated in figure 6.12. As in ethylene, the numbering of them is in order of increasing energy. In figure 6.12 the orbital symmetry species are also given: these can be obtained with the character table of the $\mathbf{C_{2h}}$ point group (to which *s-trans* butadiene belongs) given in table 4.22.

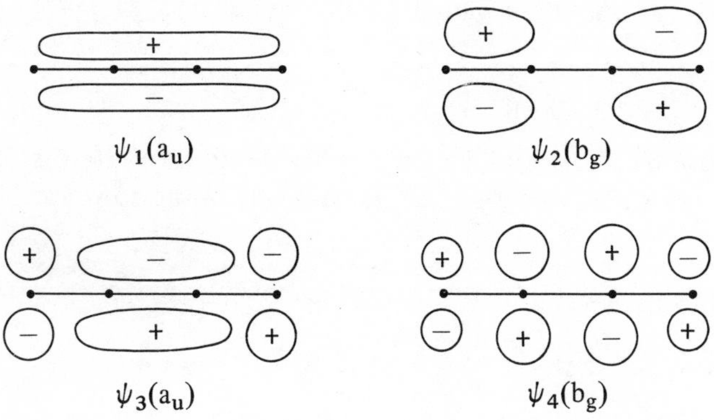

FIGURE 6.12
The four lowest energy π-MO's of butadiene

The configuration of the four π-electrons in the ground state of *s-trans* butadiene is therefore

$$(a_u)^2(b_g)^2$$

and the first excited configuration in which an electron is promoted from ψ_2 to ψ_3 is

$$(a_u)^2(b_g)(a_u)$$

The MO's of glyoxal (see figure 3.9) could be obtained in an analogous way to those of butadiene, except that different values of α and β would have to be used because of the presence of two oxygen atoms.

In the case of the cyclic conjugated system in benzene, six MO's result from a Hückel treatment of the six 2p AO's. Their energies are obtained from the secular determinant

$$\begin{vmatrix} x & 1 & 0 & 0 & 0 & 1 \\ 1 & x & 1 & 0 & 0 & 0 \\ 0 & 1 & x & 1 & 0 & 0 \\ 0 & 0 & 1 & x & 1 & 0 \\ 0 & 0 & 0 & 1 & x & 1 \\ 1 & 0 & 0 & 0 & 1 & x \end{vmatrix} = 0 \qquad (6.40)$$

and are illustrated in figure 6.13. ψ_2 and ψ_3 are doubly degenerate, that is they have equal energies, as are ψ_4 and ψ_5. The six MO wave functions are as follows

$$\begin{aligned}
\psi_1 &= \chi_1/\sqrt{6} + \chi_2/\sqrt{6} + \chi_3/\sqrt{6} + \chi_4/\sqrt{6} + \chi_5/\sqrt{6} + \chi_6/\sqrt{6} \\
\psi_2 &= \chi_2/2 + \chi_3/2 - \chi_5/2 - \chi_6/2 \\
\psi_3 &= \chi_1/\sqrt{3} + \chi_2/\sqrt{12} - \chi_3/\sqrt{12} - \chi_4/\sqrt{3} - \chi_5/\sqrt{12} + \chi_6/\sqrt{12} \\
\psi_4 &= -\chi_2/2 + \chi_3/2 - \chi_5/2 + \chi_6/2 \\
\psi_5 &= \chi_1/\sqrt{3} - \chi_2/\sqrt{12} - \chi_3/\sqrt{12} + \chi_4/\sqrt{3} - \chi_5/\sqrt{12} - \chi_6/\sqrt{12} \\
\psi_6 &= \chi_1/\sqrt{6} - \chi_2/\sqrt{6} + \chi_3/\sqrt{6} - \chi_4/\sqrt{6} + \chi_5/\sqrt{6} - \chi_6/\sqrt{6}
\end{aligned} \qquad (6.41)$$

although for the doubly degenerate orbitals other forms of the wave functions exist which have complex (as opposed to real) coefficients. The forms of the

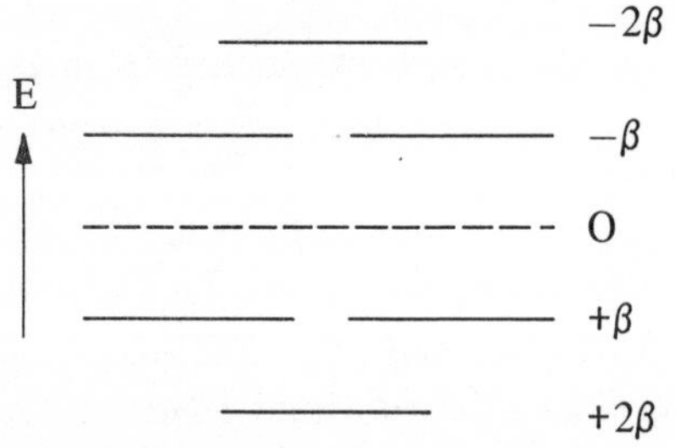

FIGURE 6.13
π-MO energy level diagram for benzene

real wave functions above the plane of the molecule are illustrated in figure 6.14. The parts below the plane are identical in form, but opposite in sign, to those above. In figure 6.14 the symmetry species assignments are also given. Benzene belongs to the $\mathbf{D}_{6h}$ point group and the assignment of the species can be confirmed using table 4.36.

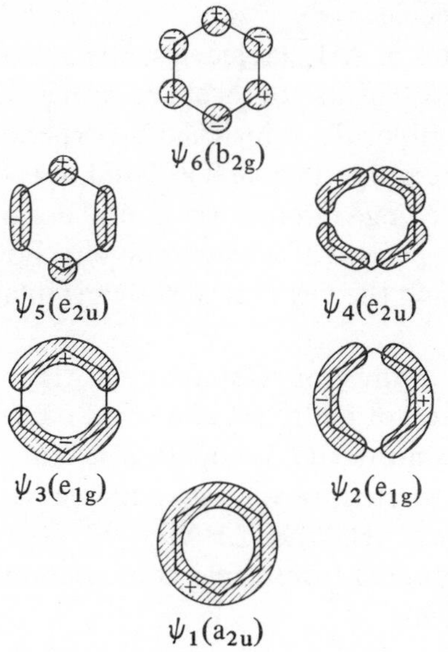

FIGURE 6.14
The six lowest energy π-MO's of benzene

The ground state configuration of the π-electrons in benzene is

$$(a_{2u})^2(e_{1g})^4$$

and the first excited configuration is

$$(a_{2u})^2(e_{1g})^3(e_{2u})$$

6.3 Walsh molecular orbital diagrams

Another group of molecules in which an approximate MO method has had considerable success consists of small molecules containing elements mainly in the first two rows of the periodic table. Orbitals of molecules with generalized formulae AH_2, AB_2, BAC, HAB, AH_3, AB_3, and H_2AB, where A, B, and C are usually first or second row elements, have been derived in a semi-quantative way with particular stress on relative energies of orbitals and the way in which they change with nuclear configuration.

Here we shall consider in detail only the AH_2 molecules and the reader is referred to the original papers (Walsh, A. D., *J. Chem. Soc.*, p. 2260 *et seq*, 1953) for the treatment of the other molecules.

In describing the ground electronic states of AH_2 molecules, MO's which are *localized* in the AH bonds are adequate. The use of localized rather than *delocalized* MO's, which cover the whole molecule, is justified by experimental observations. These show that ground state properties such as bond dissociation energies, bond lengths and bond stretching vibration wavenumbers are almost constant for a particular bond, irrespective of its environment. For example the O—H bond in water, phenol and the OH radical shows similar properties.

The localized MO's which are occupied in the ground state of an AH_2 molecule are constructed from the 1s orbital of hydrogen and s and p orbitals of A: these are 2s and 2p if A is a first-row atom and 3s and 3p if it is a second-row atom. The MO's must be constructed for the two possible extreme values of 90° and 180° for the angle HAH. When ∠HAH = 90°, the localized MO's in each AH bond are constructed from linear combinations of p orbitals on A and 1s orbitals on H giving

$$\psi_{\text{loc}} = \chi(\text{H:1s}) + \chi(\text{A:}n\text{p}) \tag{6.42}$$

where ψ_{loc} is the localized MO wave function and $n = 2$ or 3. The wave functions in each AH bond are illustrated in figure 6.15. The remaining np orbital and the ns orbital on A are virtually unchanged AO's.

For ∠HAH = 180° the orbitals on A which form AH bonds cannot be pure s or p orbitals, but must be sp hybrid orbitals of equal s and p character. The two hybrid orbitals on A are at an angle of 180° to each other, as shown in figure 6.16(a). The localized MO's are then constructed by making linear combinations of sp hybrid and hydrogen ls orbitals giving

$$\psi_{\text{loc}} = \chi(\text{H:1s}) + \chi(\text{A:sp}) \tag{6.43}$$

These localized wave functions are illustrated in figure 6.16(b). The two remaining p orbitals are virtually unchanged AO's.

However, when we consider electronic excited states a localized MO description of electron configurations is no longer adequate. Taking H_2O as an example, this has in the ground state two electrons in each of two localized MO's, similar to those of figure 6.15 (in fact ∠HOH = 104°31′). In an excited state an electron is promoted from one of the localized MO's: but from which one? The question is not a meaningful one because we must use delocalized MO's when discussing electronic excitation.

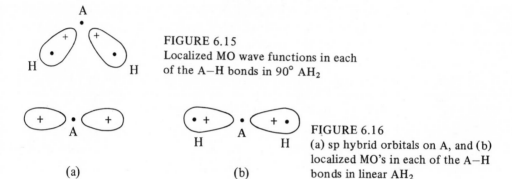

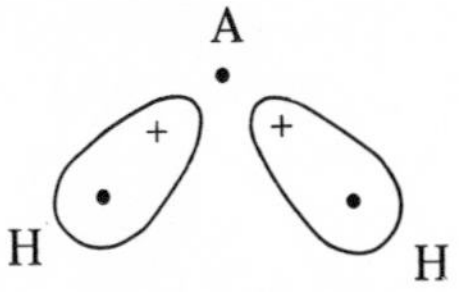

FIGURE 6.15
Localized MO wave functions in each of the A–H bonds in 90° AH_2

(a) (b)

FIGURE 6.16
(a) sp hybrid orbitals on A, and (b) localized MO's in each of the A–H bonds in linear AH_2

The delocalized MO's of AH_2, where A is a first row element, will now be discussed for the two extreme HAH angles and correlated to indicate what happens to them at intermediate angles.

(i) ∠HAH = 180°

The two localized MO's of figure 6.16(b) are delocalized simply by taking in-phase and out-of-phase combinations of them. The in-phase combination produces the MO at the lower righthand side of figure 6.17. Linear AH_2 belongs to the $\mathbf{D}_{\infty h}$ point group whose character table is given in table 4.37. This can be used to classify the MO as σ_g (in fact it is σ_g^+ but the + is usually left off the symbol). Similarly the out-of-phase combination produces the σ_u orbital illustrated. The two atomic p orbitals on A are degenerate in linear AH_2 and are both π_u. The MO's are arranged in figure 6.17 in order of decreasing 'binding energy'. This energy is taken as a measure of the strength of binding of the electrons; for example, the least strongly bound electron will be the most easily promoted to an orbital of lower binding energy, or removed in an ionization process. The general rule that more s-character in an orbital tends to lead to higher binding energy is used. This rule arises because an electron in an *ns* orbital spends more of its time close to the nucleus than if it were in an *np* orbital.

(ii) ∠HAH = 90°

The two localized MO's of figure 6.15 are delocalized by taking in-phase and out-of-phase combinations. When classified according to the $\mathbf{C}_{2v}$ point group,

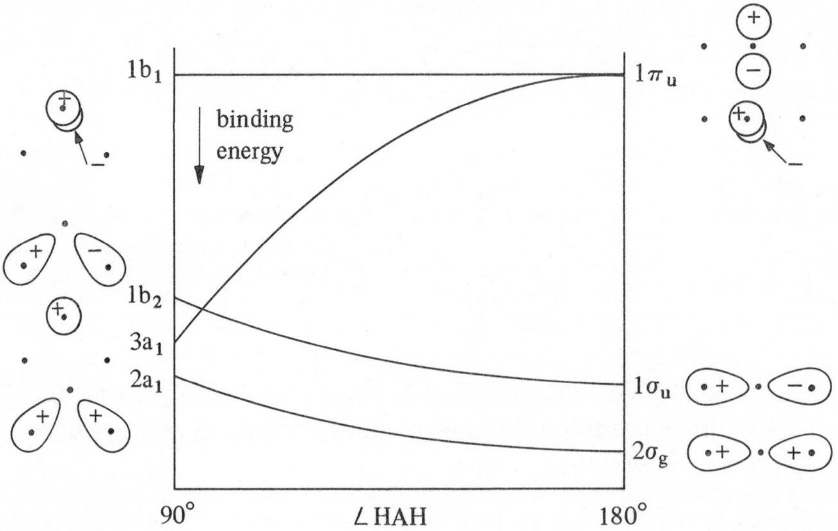

FIGURE 6.17
Walsh MO diagram for AH_2 molecules

to which bent AH_2 belongs, there is an a_1 and a b_2 orbital: these are illustrated on the lefthand side of figure 6.17. The atomic s and p orbitals on A are respectively a_1 and b_1 orbitals. When A is a first-row atom the two a_1 orbitals are distinguished by numbering them 2 and 3 in order of decreasing binding energy. (The 1s AO on A becomes $1a_1$ when ∠HAH = 90° and $1\sigma_g$ when ∠HAH = 180°).

Correlation of the orbitals as ∠HAH changes from 90° to 180° is obvious in respect of $2a_1 - 2\sigma_g$, $1b_2 - 1\sigma_u$, and $1b_1 - 1\pi_u$ but less obvious for $3a_1 - 1\pi_u$. This last correlation is especially important since the binding energy of the orbital is much larger for the 90° than the 180° angle because of the change from pure s to pure p character.

Such an MO diagram can be used to predict the HAH angle in the ground and excited states of AH_2 molecules. A few examples will now be considered.

(a) BeH_2; MgH_2

The Be atom has the electron configuration $1s^2 2s^2$ and has therefore two outer electrons to be fed into MO's. Feeding these as well as the two electrons from the hydrogen atoms into the MO's in pairs, with opposed spins, we obtain the ground state configuration

$$(2\sigma_g)^2(1\sigma_u)^2$$

and the prediction that the molecule BeH_2 is linear in this state since both orbitals have a maximum binding energy in the 180° configuration. Similarly MgH_2 should have a ground state configuration $(3\sigma_g)^2$ $(2\sigma_u)^2$, the only difference from BeH_2 being that the MO's are constructed from 3s and 3p rather than 2s and 2p AO's. MgH_2 should also be linear in its ground state, but neither this molecule nor BeH_2 is known.

The first excited state of BeH_2 will have the configuration

$$(2a_1)^2(1b_2)(3a_1)$$

and will probably be non-linear because of the effect of one electron in the $3a_1$ orbital.

(b) NH_2

The N atom has the electron configuration $(1s)^2(2s)^2(2p)^3$ and therefore five outer electrons to be fed into MO's, along with two from the hydrogen atoms. The ground state configuration is

$$(2a_1)^2(1b_2)^2(3a_1)^2(1b_1)$$

leading to a bent molecule, because the effect of two electrons in the $3a_1$ orbital should outweigh all others. The observed value of the angle in the short-lived NH_2 radical is 103°21′. The configuration of the first excited state is

$$(2a_1)^2(1b_2)^2(3a_1)(1b_1)^2$$

leading to a larger angle than in the ground state since an electron has been removed from the $3a_1$ orbital, which has a maximum binding energy at 90°, to the $1b_1$ orbital which has the same energy for all values of the angle. The observed value of the angle in the first excited state is 144°.

(c) H_2O

The O atom has the electron configuration $(1s)^2(2s)^2(2p)^4$. In H_2O there are therefore eight electrons, six from O and two from the hydrogens, to be fed into the MO's, giving the ground state configuration

$$(2a_1)^2(1b_2)^2(3a_1)^2(1b_1)^2$$

leading to a bent molecule: the observed angle is 104°31′. Some of the lower lying excited states result from promotion of an electron from the $1b_1$ MO into orbitals which are so far from the nuclei that they do not affect the geometry very much (so-called Rydberg orbitals—see section 6.4). The angle in one such excited state is known to be 106°9′ which is consistent with promotion from $1b_1$ into a Rydberg orbital.

6.4 Rydberg orbitals

In section 6.1, MO's for homonuclear diatomic molecules were built up using an LCAO treatment of the various atomic orbitals. In general these MO's have the form

$$\psi_{\pm}(ns) = N[\chi_1(ns) \pm \chi_2(ns)] \tag{6.44}$$

$$\psi_{\pm}(np) = N[\chi_1(np) \pm \chi_2(np)] \tag{6.45}$$

$$\psi_{\pm}(nd) = N[\chi_1(nd) \pm \chi_2(nd)] \tag{6.46}$$

etc., where N represents various normalization constants, the $\psi_{\pm}$ are the various MO wave functions and the χ_1 and χ_2 are AO wave functions centred on nucleus 1 and 2 respectively. Some of the lower energy MO's are illustrated in figure 6.4.

Experimentally it is found that the higher energy orbitals, given by equations 6.44-46 for high n, converge to a limit which represents, just as in an atom, the removal of an electron in an ionization process. The analogy between molecules and atoms goes closer than this: it is found that, when n is not small, the orbital energies in a molecule follow the expression

$$E_n/hc = -R/(n - \delta)^2 \tag{6.47}$$

where R is the Rydberg constant for the molecule concerned, δ is called the quantum defect and n is an integer. It is because the orbital energies follow the expression given in equation 6.47, which is very similar to equation 1.1

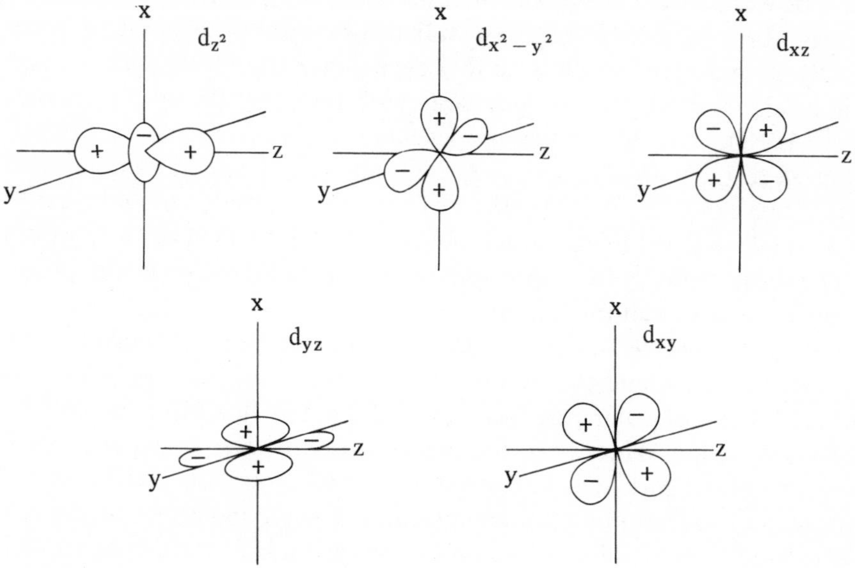

FIGURE 6.18
Illustration of the five degenerate d atomic orbitals

for atomic hydrogen and identical to the corresponding expression for polyelectronic atoms, that the orbitals are called Rydberg orbitals.

Any MO will resemble an AO when the MO becomes large compared to the size of the molecular core. For example, an electron in a high energy orbital in O_2 will tend to see the O_2^+ core as resembling the S^+ core of a sulphur atom, because the separation of the two oxygen nuclei is small compared to size of the MO and the effect of the core can be represented usefully by what is called the *united atom approximation.* Thus, in O_2 we should be able to correlate all the MO's of equations 6.44-46 with AO's of the sulphur atom and similarly for any homonuclear diatomic molecule. For low *n*, the perturbation of the united atom AO by the molecular core is large and the united atom approximation is poor, but it becomes increasingly good as *n* increases.

From figure 6.4, it is obvious that, as the internuclear distance is decreased to zero, the $\sigma_g 1s$ MO becomes, in the united atom, a 1s AO which is spherically symmetrical, the $\sigma_u 1s$ MO becomes a $2p_z$ AO and the $\sigma_g 2s$ MO becomes a 2s AO. However, the correlation of some of the higher energy MO's with united atom AO's is not quite so straightforward. For example $\sigma_g 2p$ correlates with 3s and $\pi_g 2p$, which is doubly degenerate, correlates with the two 3d orbitals $3d_{xz}$ and $3d_{yz}$ illustrated, along with the other three 3d orbitals, in figure 6.18. The five 3d orbitals are degenerate in an atom and the labels usually given to them are d_{z^2}, $d_{x^2-y^2}$, d_{xy}, d_{yz}, and d_{xz}. It should be noted that this is only one possible set of d orbitals. For example, the x-, y- and z-axis labels could be switched in any order and equally valid sets of orbitals would result. The set in figure 6.18 is appropriate here since the z-axis is unique in this set as it is also in diatomic molecules.

Table 6.1 lists some of the lower energy MO-united atom AO correlations.

Although the MO's tend to resemble AO's at higher energies there are two effects which the molecular core has on the MO's. The first effect is on the energy so that, at low values of *n*, the energies are not given by equation 6.47. However, this effect soon dies out and, for example, is usually small when $n = 3$ for second row atom diatomic molecules. The second effect of the core extends to much higher values of *n* and can be regarded as the degree to which the Rydberg orbital feels the effect of the *symmetry* of the core. The result of this perturbation by the core is that all but the highest energy Rydberg orbitals should be classified, in a homonuclear diatomic molecule, according to the $D_{\infty h}$ point group as in the lefthand column of table 6.1. In fact it is easier, from the point of view of symmetry classification, to go from the united atom AO's to the MO's than *vice versa*.

In classifying AO's in any point group there are useful general rules concerning s, p and d orbitals.

(a) An s AO always has the totally symmetric species of the point group.
(b) The p_x, p_y and p_z AO's always have the species of the translations T_x, T_y, and T_z respectively.

(c) The d_{z^2}, $d_{x^2-y^2}$, d_{xy}, d_{yz}, and d_{xz} AO's have the species of the components of the polarizability α_{z^2} (or $\alpha_{2z^2-x^2-y^2}$), $\alpha_{x^2-y^2}$, α_{xy}, α_{yz}, and α_{xz} respectively. The symmetry species of the polarizability components are given in the last column of each of the character tables in tables 4.1-45 and are discussed in section 7.5.

Using these rules and the $\mathbf{D}_{\infty h}$ character table in table 4.37 the s, p, and d orbitals can be classified in this point group: the results are given in table

TABLE 6.1
Correlation of homonuclear diatomic molecule MO's with united atom AO's

Diatomic molecule MO	*United atom AO*
$\sigma_g 1s$	1s
$\sigma_u 1s$	$2p_z$
$\sigma_g 2s$	2s
$\sigma_u 2s$	$3p_z$
$\sigma_g 2p$	3s
$\pi_u 2p$	$2p_x$, $2p_y$
$\pi_g 2p$	$3d_{xz}$, $3d_{yz}$
$\sigma_u 2p$	$4p_z$
$\sigma_g 3s$	$3d_{z^2}$
$\sigma_u 3s$	$5p_z$
$\sigma_g 3p$	4s
$\pi_u 3p$	$3p_x$, $3p_y$
$\pi_g 3p$	$4d_{xz}$, $4d_{yz}$

6.2 (usually the superscript '+' is dropped in MO symbolism). This table shows that the threefold degenerate *n*p united atom AO's are split by the $\mathbf{D}_{\infty h}$ core into a nondegenerate $\sigma_u np$ and a doubly degenerate $\pi_g np$ MO and the fivefold degenerate *n*d AO's are split into three groups of MO's. The splittings all tend to zero as *n* tends to infinity.

In polyelectronic atoms the quantum defect δ in equation 6.47 is a measure of the degree of 'penetration' of the core by the orbital concerned. This penetration is greatest for s orbitals which have a maximum in the wave function at the nucleus and consequently a large value of $\delta(\sim 1.0)$: similarly a smaller value of $\delta(\sim 0.5)$ is associated with p orbitals and a still smaller value (~ 0.05) with d orbitals. In diatomic molecules δ is also a measure of the penetration of the molecular core by the Rydberg orbital and has a different value, for example, in $\pi_g np$ and $\sigma_u np$ orbitals.

In heteronuclear diatomic molecules, such as NO and CO, in which the nuclear charges are similar, Rydberg orbitals have a very similar form to those in homonuclear diatomics and the correlations between atom AO's and MO's are similar to those in table 6.1 except that the 'g' and 'u' labels on the MO's are dropped. In a molecule like HCl, however, the united atom AO and MO correlations are quite different.

TABLE 6.2
Symmetry species in the $\mathbf{D}_{\infty h}$ point group of molecular orbitals derived from s, p, and d united atomic orbitals

United atom orbital	*Molecular orbital*
s	σ_g^+
p_z	σ_u^+
p_x, p_y	π_g
d_{z^2}	σ_g^+
d_{xz}, d_{yz}	π_g
$d_{x^2-y^2}, d_{xy}$	δ_g

In polyatomic molecules MO's again begin to resemble perturbed AO's at higher energy, but the actual correlations are more difficult to make. As an example let us consider the lower energy π orbitals of benzene, illustrated in figure 6.14. The a_{2u} orbital correlates with a p_z AO in the united atom since (see table 4.36) a_{2u} is the species of T_z. The doubly degenerate e_{1g} orbitals correlate with the degenerate d_{xz}, d_{yz} AO's in the united atom since e_{1g} is the symmetry species of $(\alpha_{xz}, \alpha_{yz})$. We have not dealt with f or g AO's but, in fact, the e_{2u} and b_{2g} π-MO's correlate with f and g AO's respectively in the united atom.

Finally, it is perhaps worth drawing attention to the value of n in equation 6.47. The value in this equation is that which labels the united atom AO's for example in the righthand column of table 6.1. This table illustrates the fact that the value of n in equation 6.47 is not usually the same as the value of n which labels the MO, as in the lefthand column of table 6.1. In homonuclear diatomic molecules the correlation in table 6.1 is easy to derive, but in other molecules it can be much more difficult.

6.5 Crystal field and ligand field molecular orbitals

Transition metal atoms are distinguished from other atoms by their highest energy occupied orbitals being d orbitals. Each of the 3d, 4d, 5d sub-shells can accommodate ten electrons so there are ten atoms in each of the first, second and third transition series. In this section we shall be considering only atoms in the first transition series namely Sc, Ti, V, Cr, Mn, Fe, Co, Ni, Cu, and Zn which have outer electrons in the 3d sub-shell.

Transition metals commonly form complex ions, for example $[Fe(CN)_6]^{4-}$, the ferrocyanide ion, and $[Fe(H_2O)_6]^{2+}$, the hydrated ferrous ion. The MO treatment of such ions has been developed separately from that of the kinds of molecules we have discussed in sections 6.1-4. It is for this reason that the expressions 'Crystal Field Theory' and 'Ligand Field Theory' have arisen, whereas it might have been less confusing if these names had not been adopted since both are simply particular aspects of MO theory.

In $[Fe(CN)_6]^{4-}$ and $[Fe(H_2O)_6]^{2+}$ it is usual to give the name *ligand* to a CN or H_2O group attached to the metal atom. The word ligand has arisen because the type of bonding between the metal atom and the groups attached to it is different from that in, say, HCl. However, since this is a quantitative rather than a qualitative difference, the description of the CN group in $[Fe(CN)_6]^{4-}$ as a ligand, but not the H in HCl, can also be rather confusing.

The ligands in a transition metal complex are usually arranged in a highly symmetrical way about the central metal atom. For example an octahedral arrangement of six ligands, as in the two complex ions already mentioned and also the ferricyanide ion $[Fe(CN)_6]^{3-}$, illustrated in figure 2.3(a), is common and it is this arrangement which we will consider in some detail.

The higher energy molecular orbitals in an octahedral transition metal complex are perturbed d orbitals of the metal atom. If the perturbation is weak the ligands can be regarded as point negative charges at the corners of a regular octahedron at the centre of which is the metal atom. The perturbation by the ligands of the d orbitals *from outside* closely resembles, in respect of the symmetry properties of the d orbitals in the perturbing environment, the perturbation by the molecular core of d orbitals *from inside* in the case of Rydberg orbitals (section 6.4). The MO theory, which is used in cases where the ligands can be treated as point charges, is called *crystal field theory.* The name arises because it was originally developed by Bethe for the treatment of modification of orbital energies of a free ion by other ions surrounding it in a crystal, for example the perturbation of the orbital energies of Na^+ when surrounded, in the NaCl crystal, by octahedrally arranged nearest-neighbour Cl^- ions.

If the ligands interact more strongly with the metal atom then the ligand cannot be treated as a point charge, but as a group which, in isolation, has its own MO's which will interact with the d orbitals on the metal atom. The MO theory used in these cases is called *ligand field theory.*

6.5.1 CRYSTAL FIELD THEORY

In an octahedral field due to six point charges the d orbitals of the central atom have to be reclassified according to the O_h point group (see table 4.43 for character table). With the set of d orbitals illustrated in figure 6.18 we place the six charges on the cartesian axes such that the metal atom is at the origin.

As was mentioned in section 6.4 the d orbitals always have the symmetry species of the corresponding components of the polarizability in the point group corresponding to the perturbation. In the case of an octahedral field the symmetry species of the d_{z^2}, $d_{x^2-y^2}$, d_{xy}, d_{yz}, and d_{xz} orbitals are respectively those of the components of the polarizability $\alpha_{2z^2-x^2-y^2}$, $\alpha_{x^2-y^2}$, α_{xy}, α_{yz}, and α_{xz}: from table 4.43 we see that the symmetry species are e_g, e_g, t_{2g}, t_{2g}, and t_{2g} respectively. So the fivefold degenerate d orbitals are split in an octahedral field into a doubly degenerate and a triply degenerate set of what are now molecular orbitals (the degeneracy of the d_{xy}, d_{yz}, and d_{xz} orbitals in an octahedral field is obvious, but this is not so in the case of the other two orbitals). Whether the e_g orbitals are at higher or lower energy than the t_{2g} orbitals is determined by electron repulsions. The d_{z^2} and $d_{x^2-y^2}$ orbitals have much of the associated electron density in the line of the metal-ligand bond and therefore electrons in these orbitals experience more repulsion from the electrons of the ligands than would electrons in the d_{xy}, d_{yz}, or d_{xz} orbitals. The e_g orbitals are therefore higher in energy than the t_{2g} orbitals.

The splitting between the e_g and t_{2g} orbitals is usually labelled Δ and is shown in figure 6.19. The value of Δ is often such that the promotion of an electron from a t_{2g} to an e_g orbital results in an absorption spectrum in the visible region leading to the characteristic property of transition metal complexes that they are highly coloured.

Now let us consider how the 1, 2, 3 . . . 10 electrons of a transition metal atom can be fed into the MO's in figure 6.19 to give the ground state configuration. In the case of one electron this goes into any one of the t_{2g} orbitals: this is shown in figure 6.20 in which the three t_{2g} MO's have been separated for convenience. For two electrons, they will each go into different t_{2g} MO's with their spins parallel according to one of Hund's rules (similar to the ground state of oxygen in section 6.1) rather than into the same t_{2g} orbital

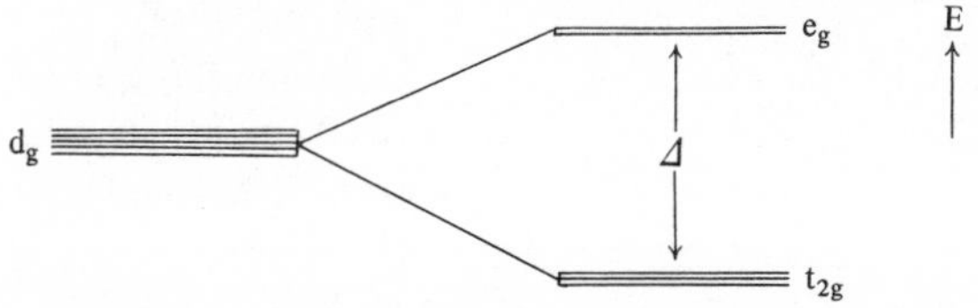

FIGURE 6.19
The splitting of d orbitals in an octahedral field

with their spins antiparallel: similarly three electrons will each go into different t_{2g} MO's with their spins parallel. For eight, nine and ten electrons the ground state configurations are obtained in a similar way and are also given in figure 6.20.

In the cases of ground states in which there are four, five, six or seven d electrons a complication arises. If the value of Δ is small it may be energetically favourable for electrons to be excited from a t_{2g} to an e_g MO rather than suffer inter-electron repulsions which are strongest when the electrons are in MO's in pairs, with their spins antiparallel. Δ is small when the perturbing field is low and, in these conditions, the number of unpaired spins may be high: the resulting ground configuration is called a 'weak field, high spin' configuration. For example, if there are five electrons to be fed into the MO's in a weak field, high spin configuration they will go into each of the five MO's with their spins parallel. This and other weak field, high spin configurations are shown in figure 6.21. If, on the other hand, the field is stronger the value of Δ is larger and it may be energetically favourable for the electrons to have their spins paired and suffer repulsions rather than be excited to the e_g orbitals. The ground configurations resulting under these conditions are called 'strong field, low spin' configurations and these are illustrated in figure 6.22. It should be noted that, even in the low spin case, when there are four electrons two of them go into

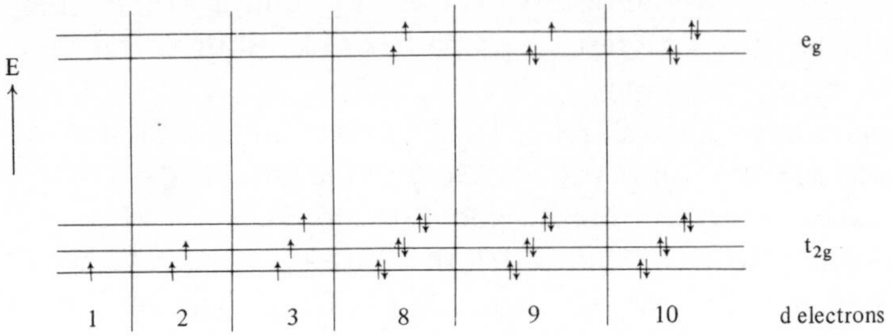

FIG. 6.20
Configurations arising in an octahedral field from 1, 2, 3, 8, 9 and 10 d electrons

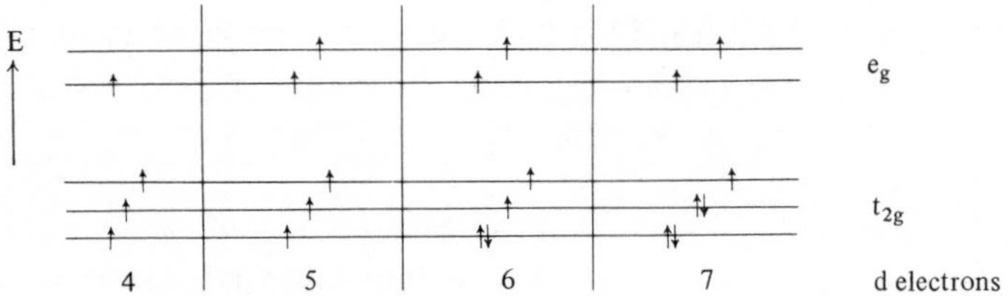

FIGURE 6.21
Configurations, in a weak octahedral field, arising from 4, 5, 6 and 7 d electrons

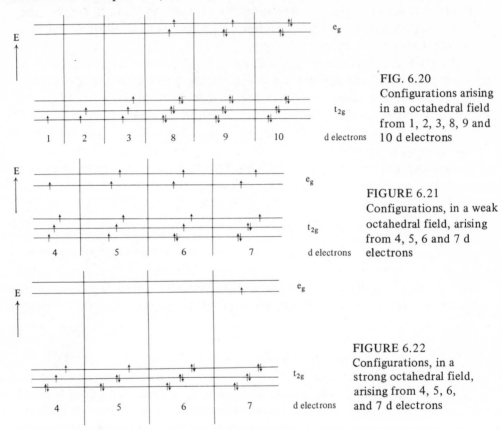

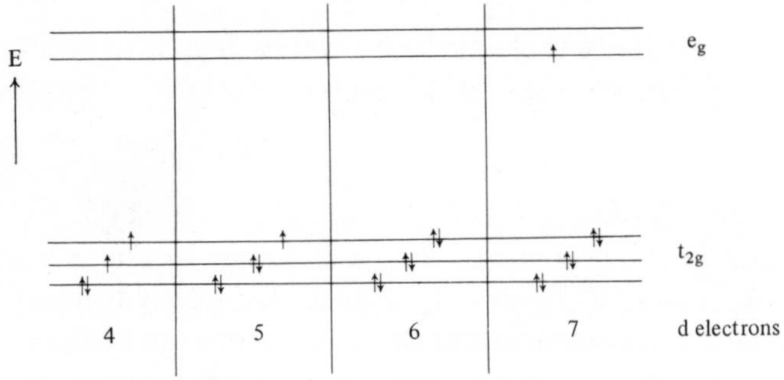

FIGURE 6.22
Configurations, in a strong octahedral field, arising from 4, 5, 6, and 7 d electrons

different t_{2g} orbitals giving the highest possible multiplicity in accordance with Hund's rule.

The weak field, high spin and strong field, low spin configurations are two extreme possibilities. In practice configurations intermediate between these extremes are found when the field is intermediate in strength.

It should be obvious now why the two weak field and strong field cases do not arise when there are other than four, five, six, or seven d electrons.

So far we have considered only an octahedral arrangement of ligands. The symmetry species of the d orbitals in fields of any symmetry can be obtained from the species of the corresponding components of the polarizability given in the appropriate character table. The d orbital species in some of the more common types of field are given in table 6.3. Some examples of complexes

TABLE 6.3
Splitting of d-orbitals by perturbing fields of various symmetries

Point group	*d-orbitals* d_{z^2}	$d_{x^2-y^2}$	d_{xy}	d_{yz}	d_{xz}
$\mathbf{O_h}$		e_g		t_{2g}	
$\mathbf{T_d}$		e		t_2	
$\mathbf{D_{3h}}$	a_1'	e'		e''	
$\mathbf{D_{4h}}$	a_{1g}	b_{1g}	b_{2g}	e_g	
$\mathbf{D_{\infty h}}$	σ_g^+	δ_g		π_g	
$\mathbf{C_{2v}}$	a_1	a_1	a_2	b_2	b_1
$\mathbf{C_{3v}}$	a_1	a_1	a_2	e	
$\mathbf{C_{4v}}$	a_1	b_1	b_2	e	
$\mathbf{D_{2d}}$	a_1	b_1	b_2	e	
$\mathbf{D_{4d}}$	a_1	e_2		e_3	

which have symmetrical arrangements of ligands are $[CrO_4]^{2-}$, tetrahedral ($\mathbf{T_d}$); $[PtCl_4]^{2-}$, square planar ($\mathbf{D_{4h}}$); $[CuCl_2]^-$, linear ($\mathbf{D_{\infty h}}$).

6.5.2 LIGAND FIELD THEORY

In crystal field theory the model treats the ligands as point charges and the ligand orbitals are disregarded. This approximation breaks down in cases where the ligand interacts more strongly with the central atom. Under these circumstances the MO's of the ligands must be taken into account in deriving the MO's for the whole complex molecule or ion.

In treating the MO's of the ligand it is convenient to separate them into two kinds (a) those which are σ orbitals with respect to the metal-ligand bond, and (b) those which are π orbitals with respect to the bond. (The σ and π labels used here are not strictly symmetry labels except in so far as the localized metal-ligand bond may be treated like that in a diatomic molecule.)

σ-Bonding between the metal and the ligand is invariably much stronger than π-bonding. Consequently, if both types occur the effect of the π-bonding, is such as to modify relatively slightly the MO's obtained by taking into account σ-bonding only. We shall not consider here the effects of π-bonding, but only the stronger effect of σ-bonding and, as in crystal field theory, we shall consider only the octahedral case in detail.

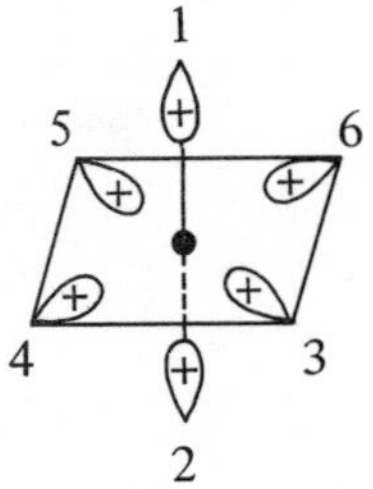

FIGURE 6.23
Six equivalent ligand σ orbitals arranged octahedrally about a central atom

In figure 6.23 are shown the six localized σ orbitals arranged octahedrally about the metal atom. These orbitals might be, for example, sp hybrid orbitals in the case of the $(CN)^-$ ligand. In order to classify these orbitals in the $\mathbf{O_h}$ point group they have to be delocalized. In the case of two equivalent localized MO's delocalization is easy: we simply take in-phase and out-of-phase combinations (as for AH_2 molecules in section 6.3): but, for more than two, delocalization is more difficult and only the results will be presented here. In table 6.4 the symmetry species of the delocalized σ orbitals are given for various point groups. In many point groups there is more than one arrangement of ligands which can produce the same symmetry: for example the square planar arrangement of ligands in the molecule ML_4 (where M is a metal atom and L is a ligand) belongs to the $\mathbf{D_{4h}}$ point group as also does the octahedral arrangement in $ML_4L'_2$ where the two L′ ligands are *trans* to each other.

Consequently, in table 6.4 there may be more than one set of ligand σ orbital symmetry species for any one point group.

In the case of the $\mathbf{O_h}$ point group there are six ligand σ orbitals of species a_{1g}, e_g (doubly degenerate), and t_{1u} (triply degenerate). These are shown on the righthand side of figure 6.24 and are probably lower in energy than the

TABLE 6.4
Classification of σ ligand orbitals in various point groups

Point group	*σ orbital symmetry species*
$\mathbf{O_h}$	$a_{1g} + e_g + t_{1u}$ (in octahedral ML_6)
$\mathbf{T_d}$	$a_1 + t_2$ (in tetrahedral ML_4)
$\mathbf{D_{3h}}$	$2a_1' + e'$ (in trigonal bipyramidal ML_5)
$\mathbf{D_{4h}}$	$a_{1g} + b_{1g} + e_u$ (in square planar ML_4)
	$2a_{1g} + a_{2u} + b_{1g} + e_u$ (in *trans*-octahedral ML_4L_2')
$\mathbf{D_{\infty h}}$	$\sigma_g + \sigma_u$ (in linear ML_2)
$\mathbf{C_{2v}}$	$a_1 + b_2$ (in non-linear ML_2)
	$2a_1 + b_1 + b_2$ (in tetrahedral ML_2L_2')
	$3a_1 + a_2 + b_1 + b_2$ (in *cis*-octahedral ML_4L_2')
$\mathbf{C_{3v}}$	$2a_1 + e$ (in tetrahedral ML_3L')
	$2a_1 + 2e$ (in all-*cis*-octahedral ML_3L_3')
$\mathbf{C_{4v}}$	$2a_1 + b_1 + e$ (in square pyramidal ML_4L')
	$3a_1 + b_1 + e$ (in octahedral ML_5L')
$\mathbf{D_{2d}}$	$2a_1 + 2b_2 + 2e$ (in dodecahedral ML_8)
$\mathbf{D_{4d}}$	$a_1 + b_2 + e_1 + e_2 + e_3$ (in square antiprism ML_8)

d orbitals of the metal atom on the lefthand side. In the case of weak (crystal field) interaction the d orbitals are split as shown here and in figure 6.19. When interaction occurs between the ligand σ orbitals and the split d orbitals the effect is to increase the value of Δ, as shown in the centre of figure 6.24. The reason for this is as follows. The MO's, in the case of strong interaction between ligand σ orbitals and metal d orbitals, result from resonance between the localized σ orbitals and metal d orbitals. The conditions for appreciable resonance, just as in the LCAO method in diatomic molecules, are (a) that the orbitals are of comparable energy and (b) that they are of the same symmetry. The first requirement is satisfied by all the σ orbitals and the d orbitals but the second is satisfied only by the e_g metal and ligand orbitals. Consequently resonance between these pushes the e_g metal orbitals up in energy and the e_g ligand orbitals down as shown in figure 6.24 and the t_{2g}

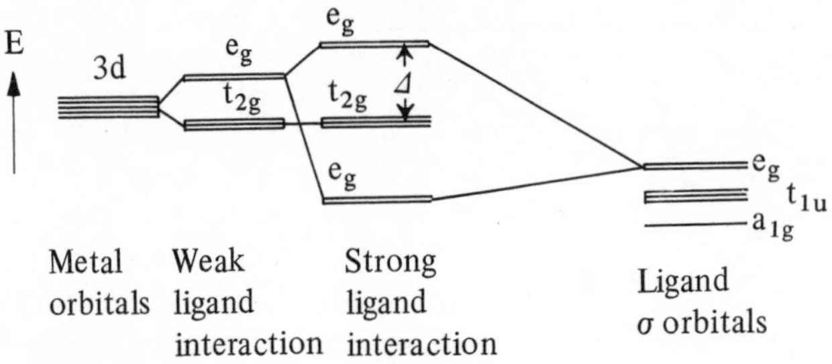

FIGURE 6.24
Molecular orbital diagram for the interaction of the d orbitals of the metal atom with the σ orbitals of octahedrally arranged ligands

metal orbitals are unaffected. It is for this reason that Δ is increased by strong interaction between the ligands and the metal.

6.6 Electronic state symmetries from orbital symmetries

In the first three sections of this chapter we have seen how it is possible in principle to calculate the MO's and hence the electron configuration of some types of molecules in their ground and some low-lying excited states.

The total electronic wave function of a molecule with a particular electron configuration has one of the symmetry species of the point group to which the molecule belongs. This symmetry species is obtained, in a non-degenerate point group, by multiplying together the symmetry species of all the occupied one-electron orbitals. (This follows from the fact that the total electronic wave function is given by the product of the wave functions of all filled one-electron MO's). For example, the symmetry species of the electronic wave function of H_2O in its ground state configuration $(2a_1)^2(1b_2)^2(3a_1)^2(1b_1)^2$

is given by the product $a_1^2 \times b_2^2 \times a_1^2 \times b_1^2$. Since the square of any species in a non-degenerate point group is the totally symmetric species we have

$$a_1^2 \times b_2^2 \times a_1^2 \times b_1^2 = A_1$$

and therefore the total electronic wave function in the ground electronic state of H_2O is A_1. (We should, of course, include also the 1s AO of oxygen which has two electrons in it, but any filled inner shell only multiplies the state species by A_1 which has no effect).

The electron spin multiplicity is also included in the symbol for an electronic state. According to the Pauli principle the two electrons in each MO must have their spins opposed so that the total spin quantum number S for all filled MO's is zero. Since $S = 0$, the multiplicity, $2S + 1$, of the ground state of H_2O is 1. The multiplicity of a state is indicated by a superscript on the symmetry species label: thus the ground state of H_2O is $^1A_1$. Similarly, the ground state of any closed shell molecule (i.e. having no partly-filled orbitals) belonging to a non-degenerate point group will always be a totally symmetric singlet state. Other examples are naphthalene ($\mathbf{D}_{2h}$ point group) in which the ground state is ^1A_g and the short-lived molecule HNO ($\mathbf{C}_s$ point group) in which the ground state is $^1A'$.

In open shell molecules belonging to non-degenerate point groups the ground state is not, in general, totally symmetric, but the symmetry species can again be obtained by multiplying the species of all the occupied orbitals. For example NH_2 (section 6.3) has the ground state configuration.

$$(2a_1)^2(1b_2)^2(3a_1)^2(1b_1)$$

The product of all these symmetry species is B_1. Because there is a single electron in the $1b_1$ orbital, $S = \frac{1}{2}$, $2S + 1 = 2$ and the ground state is therefore a doublet state and is denoted by $^2B_1$.

Another example of an open shell molecule is the bent radical BH_2 whose ground state configuration is

$$(2a_1)^2(1b_2)^2(3a_1)$$

and it is therefore a $^2A_1$ state.

Symmetry species of excited electronic states are obtained in an analogous way to those of ground states by multiplying the species of occupied orbitals.

For example, the lowest excited state of ethylene (**D_{2h}** point group), considering, as in the Hückel treatment (section 6.2), only the π-electrons, has the configuration

$$(b_{3u})(b_{2g})$$

and, since $b_{3u} \times b_{2g} = B_{1u}$, this excited state is a B_{1u} state. Because the electron promoted to the b_{2g} orbital may, without violating the Pauli principle, have its spin parallel or anti-parallel to that of the electron in the b_{3u} orbital, S can be 0 or 1, giving rise to a singlet and a triplet state with the above configuration: these states are $^1B_{1u}$ and $^3B_{1u}$. One of Hund's rules says that, of the states which arise from a particular electron configuration, those of highest multiplicity lie lowest in energy. Therefore in ethylene the $^3B_{1u}$ excited state is lower in energy than $^1B_{1u}$.

The states arising from the lowest excited π-electron configuration in butadiene which is $(a_u)^2(b_g)(a_u)$, as derived in section 6.2, are ^1B_u and ^3B_u since $b_g \times a_u = B_u$.

The lowest excited configuration of NH_2, $(2a_1)^2(1b_2)^2(3a_1)(1b_1)^2$ as derived in section 6.3, arises from promotion of an electron from the $3a_1$ to the $1b_1$ orbital and gives only one state, namely $^2A_1$.

Formaldehyde ($H_2C = 0$, figure 6.25) belongs to the C_{2v} point group and is isoelectronic with, that is has the same number of electrons as, ethylene ($H_2C = CH_2$). The electronic structure of the double bond is similar to that in ethylene, but the oxygen atom has two electrons in an orbital which is non-bonding (n) and which can be regarded as a $2p_y$ AO (figure 6.25(a)): this orbital has the symmetry species b_2 and is the highest energy occupied orbital. The lowest energy unoccupied orbital is an antibonding π orbital (π^*) of species b_1 (figure 6.25(b)) which is similar to the lowest unoccupied orbital in ethylene. The lowest energy excited state results from a $b_1 - b_2$ ($\pi^* - n$) electron promotion and the symmetry species of the state is given by

$$b_1 \times b_2 = A_2$$

Therefore a singlet and a triplet state, $^1A_2$ and $^3A_2$, result from this promotion and the triplet is lower in energy.

In a degenerate point group the problem of obtaining the symmetry species of the total electronic wave function corresponding to a particular configuration is rather more difficult. For example, the ground state configuration of the π-electrons in benzene was shown in section 6.2 to be $(a_{2u})^2(e_{1g})^4$. Since $a_{2u}{}^2 = {}^1A_{1g}$, the ground state symmetry species is obtained from $(e_{1g})^4$. Even taking into account the requirement that the spins of the four electrons in the e_{1g} orbitals must be paired which allows only singlet states, the conclusions of section 4.3.3 might lead us to expect several states to arise from the $(e_{1g})^4$ configuration. In fact only one of the states, ${}^1A_{1g}$, is allowed: this can be shown by a rather complex application of the Pauli principle. It turns out that the symmetry species of the ground electronic state of *any* closed shell molecule is totally symmetric irrespective of whether the molecule belongs to a degenerate or a non-degenerate point group.

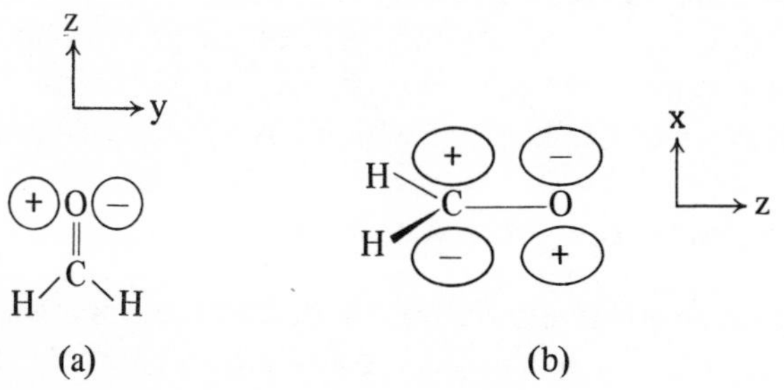

FIGURE 6.25
(a) n and (b) π^* MO's of formaldehyde

The symmetry species of the ground state electronic wave function of an open shell molecule belonging to a degenerate point group is easy to obtain if the open shell has only one electron or one vacancy. A shell with one vacancy can be treated in the same way as a shell with one electron in it. For example, the ground state configuration of the benzene positive ion $C_6H_6{}^+$ is $(a_{2u})^2(e_{1g})^3$. Since electrons and vacancies can be regarded as equivalent this configuration can be treated in the same way as $(a_{2u})^2(e_{1g})$ from which the only state arising is ${}^2E_{1g}$. Therefore the ground state of $C_6H_6{}^+$ is ${}^2E_{1g}$.

In an open shell molecule with two vacancies in a doubly degenerate orbital for example $C_6H_6{}^{2+}$ with the ground configuration $(a_{2u})^2(e_{1g})^2$, more than one electronic state arises from the configuration. The states which arise from the $(e_{1g})^2$ configuration can be obtained using table 4.52 which gives the result that

$$e_{1g} \times e_{1g} = A_{1g} + A_{2g} + E_{2g}$$

The electron spin necessitates the use of $e_{1g} \times e_{1g}$ *in deriving the states rather than* $(e_{1g})^2$. With two electrons in a doubly degenerate orbital there may be singlet or triplet states but application of the Pauli principle shows that the only allowed states are ${}^1A_{1g}$, ${}^3A_{2g}$, and ${}^1E_{2g}$ i.e. the symmetric part (section 4.3.3) of the product gives the singlet states and the antisymmetric part the triplet state. The three states have different energies but according to Hund's rule ${}^3A_{2g}$ will be the ground state. The order of the other two states cannot be predicted easily.

Diatomic and linear polyatomic molecules belong to either of the degenerate point groups $\mathbf{D_{\infty h}}$ or $\mathbf{C_{\infty v}}$. If they are closed shell molecules the ground state symmetry species is the totally symmetric one, namely ${}^1\Sigma_g^+$ or ${}^1\Sigma^+$. For example the ground states of N_2 and CO both of which have the configurations

$$(\sigma 1s)^2(\sigma^* 1s)^2(\sigma 2s)^2(\sigma^* 2s)^2(\pi 2p)^4(\sigma 2p)^2$$

have ${}^1\Sigma_g^+$ and ${}^1\Sigma^+$ ground states respectively. Nitric oxide has the open shell ground state configuration . . . $(\pi 2p)^4(\sigma 2p)^2(\pi^* 2p)$ so the ground state is ${}^2\Pi$. Oxygen has the ground open shell configuration . . . $(\sigma_g 2p)^2(\pi_g^* 2p)^2$: the states which arise can be obtained using table 4.52 which shows that

$$\pi_g \times \pi_g = \Sigma_g^+ + \Sigma_g^- + \Delta_g$$

The symmetric part of this product is $\Sigma_g^+ + \Delta_g$ and the antisymmetric part is Σ_g^- so the resulting states are ${}^1\Sigma_g^+$, ${}^3\Sigma_g^-$, and ${}^1\Delta_g$. It has been shown experimentally that of these ${}^3\Sigma_g^-$ is the ground state and ${}^1\Delta_g$ the lowest excited state: ${}^1\Sigma_g^+$ is a higher energy excited state.

Table 6.5 lists the states arising from ground configurations in molecules belonging to degenerate point groups and having degenerate orbitals which contain more than one electron or vacancy.

Excited state symmetry species in molecules belonging to degenerate point groups are obtained in a way analogous to that used for ground states. For example the first excited configuration of benzene is (section 6.2)

TABLE 6.5
States arising from ground configurations in which a degenerate orbital has more than one electron or vacancy

Point group	*Configuration*	*States*
C_3	$(e)^2$	$^3A + {}^1A + {}^1E$
C_4	$(e)^2$	$^3A + {}^1A + {}^1B + {}^1B$
C_5	$(e_1)^2$	$^3A + {}^1A + {}^1E_2$
	$(e_2)^2$	$^3A + {}^1A + {}^1E_1$
C_6	$(e_1)^2$ and $(e_2)^2$	$^3A + {}^1A + {}^1E_2$
C_7	$(e_1)^2$	$^3A + {}^1A + {}^1E_2$
	$(e_2)^2$	$^3A + {}^1A + {}^1E_3$
	$(e_3)^2$	$^3A + {}^1A + {}^1E_1$
C_8	$(e_1)^2$ and $(e_3)^2$	$^3A + {}^1A + {}^1E_2$
	$(e_2)^2$	$^3A + {}^1A + {}^1B + {}^1B$
C_{3v}	$(e)^2$	$^3A_2 + {}^1A_1 + {}^1E$
C_{4v}	$(e)^2$	$^3A_2 + {}^1A_1 + {}^1B_1 + {}^1B_2$
C_{5v}	$(e_1)^2$	$^3A_2 + {}^1A_1 + {}^1E_2$
	$(e_2)^2$	$^3A_2 + {}^1A_1 + {}^1E_1$
C_{6v}	$(e_1)^2$ and $(e_2)^2$	$^3A_2 + {}^1A_1 + {}^1E_2$
$C_{\infty v}$	$(\pi)^2$	$^1\Sigma^+ + {}^3\Sigma^- + {}^1\Delta$
	$(\delta)^2$	$^1\Sigma^+ + {}^3\Sigma^- + {}^1\Gamma$
D_3	$(e)^2$	$^3A_2 + {}^1A_1 + {}^1E$
D_4	$(e)^2$	$^3A_2 + {}^1A_1 + {}^1B_1 + {}^1B_2$
D_5	$(e_1)^2$	$^3A_2 + {}^1A_1 + {}^1E_2$
	$(e_2)^2$	$^3A_2 + {}^1A_1 + {}^1E_1$
D_6	$(e_1)^2$ and $(e_2)^2$	$^3A_2 + {}^1A_1 + {}^1E_2$
C_{3h}	$(e')^2$ and $(e'')^2$	$^3A' + {}^1A' + {}^1E'$
C_{4h}	$(e_g)^2$ and $(e_u)^2$	$^3A_g + {}^1A_g + {}^1B_g + {}^1B_g$
C_{5h}	$(e_1')^2$ and $(e_1'')^2$	$^3A' + {}^1A' + {}^1E_2'$
	$(e_2')^2$ and $(e_2'')^2$	$^3A' + {}^1A' + {}^1E_1'$

TABLE 6.5 continued

Point group	*Configuration*	*States*
$\mathbf{C_{6h}}$	$(e_{1g})^2$, $(e_{1u})^2$, $(e_{2g})^2$ and $(e_{2u})^2$	${}^3A_g + {}^1A_g + {}^1E_{2g}$
$\mathbf{D_{2d}}$	$(e)^2$	${}^3A_2 + {}^1A_1 + {}^1B_1 + {}^1B_2$
$\mathbf{D_{3d}}$	$(e_g)^2$ and $(e_u)^2$	${}^3A_{2g} + {}^1A_{1g} + {}^1E_g$
$\mathbf{D_{4d}}$	$(e_1)^2$ and $(e_3)^2$	${}^3A_2 + {}^1A_1 + {}^1E_2$
	$(e_2)^2$	${}^3A_2 + {}^1A_1 + {}^1B_1 + {}^1B_2$
$\mathbf{D_{5d}}$	$(e_{1g})^2$ and $(e_{1u})^2$	${}^3A_{2g} + {}^1A_{1g} + {}^1E_{2g}$
	$(e_{2g})^2$ and $(e_{2u})^2$	${}^3A_{2g} + {}^1A_{1g} + {}^1E_{1g}$
$\mathbf{D_{6d}}$	$(e_1)^2$ and $(e_5)^2$	${}^3A_2 + {}^1A_1 + {}^1E_2$
	$(e_2)^2$ and $(e_4)^2$	${}^3A_2 + {}^1A_1 + {}^1E_4$
	$(e_3)^2$	${}^3A_2 + {}^1A_1 + {}^1B_1 + {}^1B_2$
$\mathbf{D_{3h}}$	$(e')^2$ and $(e'')^2$	${}^3A_2' + {}^1A_1' + {}^1E'$
$\mathbf{D_{4h}}$	$(e_g)^2$ and $(e_u)^2$	${}^3A_{2g} + {}^1A_{1g} + {}^1B_{1g} + {}^1B_{2g}$
$\mathbf{D_{5h}}$	$(e_1')^2$ and $(e_1'')^2$	${}^3A_2' + {}^1A_1' + {}^1E_2'$
	$(e_2')^2$ and $(e_2'')^2$	${}^3A_2' + {}^1A_1 + {}^1E_1'$
$\mathbf{D_{6h}}$	$(e_{1g})^2$, $(e_{1u})^2$, $(e_{2g})^2$ and $(e_{2u})^2$	${}^3A_{2g} + {}^1A_{1g} + {}^1E_{2g}$
$\mathbf{D_{\infty h}}$	$(\pi_g)^2$ and $(\pi_u)^2$	${}^3\Sigma_g^- + {}^1\Sigma_g^+ + {}^1\Delta_g$
	$(\delta_g)^2$ and $(\delta_u)^2$	${}^3\Sigma_g^- + {}^1\Sigma_g^+ + {}^1\Gamma_g$
$\mathbf{S_4}$	$(e)^2$	${}^3A + {}^1A + {}^1B + {}^1B$
$\mathbf{S_6}$	$(e_g)^2$ and $(e_u)^2$	${}^3A_g + {}^1A_g + {}^1E_g$
$\mathbf{S_8}$	$(e_1)^2$ and $(e_3)^2$	${}^3A + {}^1A + {}^1E_2$
	$(e_2)^2$	${}^3A + {}^1A + {}^1B + {}^1B$
$\mathbf{T_d}$	$(t_1)^2$, $(t_1)^4$, $(t_2)^2$ and $(t_2)^4$	${}^3T_1 + {}^1A_1 + {}^1E + {}^1T_2$
	$(t_1)^3$	${}^4A_1 + {}^2E + {}^2T_1 + {}^2T_2$
	$(t_2)^3$	${}^4A_2 + {}^2E + {}^2T_1 + {}^2T_2$
	$(e)^2$	${}^3A_2 + {}^1A_1 + {}^1E$

TABLE 6.5 continued

Point group	*Configuration*	*States*
T	$(t)^2$ and $(t)^4$	$^3T + {}^1A + {}^1E + {}^1T$
	$(t)^3$	$^4A + {}^2E + {}^2T + {}^2T$
	$(e)^2$	$^3A + {}^1A + {}^1E$
O_h	$(t_{1g})^2$, $(t_{1u})^2$, $(t_{1g})^4$, $(t_{1u})^4$, $(t_{2g})^2$, $(t_{2u})^2$, $(t_{2g})^4$ and $(t_{2u})^4$	$^3T_{1g} + {}^1A_{1g} + {}^1E_g + {}^1T_{2g}$
	$(t_{1g})^3$	$^4A_{1g} + {}^2E_g + {}^2T_{1g} + {}^2T_{2g}$
	$(t_{1u})^3$	$^4A_{1u} + {}^2E_u + {}^2T_{1u} + {}^2T_{2u}$
	$(t_{2g})^3$	$^4A_{2g} + {}^2E_g + {}^2T_{1g} + {}^2T_{2g}$
	$(t_{2u})^3$	$^4A_{2u} + {}^2E_u + {}^2T_{1u} + {}^2T_{2u}$
	$(e_g)^2$ and $(e_u)^2$	$^3A_{2g} + {}^1A_{1g} + {}^1E_g$
O	$(t_1)^2$, $(t_1)^4$, $(t_2)^2$ and $(t_2)^4$	$^3T_1 + {}^1A_1 + {}^1E + {}^1T_2$
	$(t_1)^3$	$^4A_1 + {}^2E + {}^2T_1 + {}^2T_2$
	$(t_2)^3$	$^4A_2 + {}^2E + {}^2T_1 + {}^2T_2$
	$(e)^2$	$^3A_2 + {}^1A_1 + {}^1E$

$(a_{2u})^2(e_{1g})^3(e_{2u})$ and the multiplication of these symmetry species is equivalent to $e_{1g} \times e_{2u}$. Now table 4.52 gives the result

$$e_{1g} \times e_{2u} = B_{1u} + B_{2u} + E_{1u}$$

When two electrons (or in this case an electron and a vacancy) are in two different orbitals the Pauli principle does not exclude any of the resulting states so this configuration produces the six states $^3B_{1u}$, $^1B_{1u}$, $^3B_{2u}$, $^1B_{2u}$, $^3E_{1u}$, $^1E_{1u}$. The only simple rule regarding the ordering of these states is one of Hund's rules which results in each triplet state being lower in energy than the corresponding singlet state. It is known, however, that the lowest singlet excited state is $^1B_{2u}$ and that the lowest triplet excited state is $^3B_{1u}$ which is lower in energy than $^1B_{2u}$.

The first excited configuration of an open shell molecule belonging to a degenerate point group, for example the first excited configuration $(a_{2u})^2(e_{1g})^2(e_{2u})$ of $C_6H_6^+$, can produce several states. In this case

$$e_{1g} \times e_{1g} = {}^3A_{2g} + {}^1A_{1g} + {}^1E_{2g}$$

from table 6.5, and therefore, from table 4.52

$$\begin{aligned} e_{2u} \times (e_{1g} \times e_{1g}) &= e_{2u} \times ({}^3A_{2g} + {}^1A_{1g} + {}^1E_{2g}) \\ &= {}^4E_{2u} + {}^2E_{2u} + {}^2E_{2u} + {}^2A_{1u} + {}^2A_{2u} + {}^2E_{2u} \end{aligned}$$

remembering that the spin of 1 in the $^3A_{2g}$ state can combine with the spin of $\frac{1}{2}$ for the electron in the e_{2u} orbital to give a total spin of $\frac{1}{2}$ or $\frac{3}{2}$.

Excited configurations in diatomic and linear polyatomic molecules may also produce more than one state. For example, the first excited configuration of O_2 is . . . $(\sigma_g 2p)^2(\pi_u 2p)^3(\pi_g^* 2p)^3$ since in this molecule the order of the $\sigma_g 2p$ and $\pi_u 2p$ orbitals is switched from that given in figure 6.5: the states from this configuration are given by

$$\pi_u \times \pi_g = {}^1\Sigma_u^+ + {}^1\Sigma_u^- + {}^1\Delta_u + {}^3\Sigma_u^+ + {}^3\Sigma_u^- + {}^3\Delta_u$$

as obtained from table 4.52.

The first excited configuration of NO is . . . $(\sigma 2p)(\pi^* 2p)^2$. Now $\pi \times \pi$ gives $^1\Sigma^+ + {}^3\Sigma^- + {}^1\Delta$ (from table 6.5) and therefore

$$\sigma \times (\pi \times \pi) = {}^2\Sigma^+ + {}^2\Sigma^- + {}^4\Sigma^- + {}^2\Delta$$

N_2 has the first excited configuration . . . $(\sigma_g 2p)(\pi_g^* 2p)$ which gives only two states $^1\Pi_g$ and $^3\Pi_g$. Similarly CO, being isoelectronic with N_2, has a similar first excited configuration giving $^1\Pi$ and $^3\Pi$ states.

Further applications of molecular symmetry

7

7.1 Woodward-Hoffmann rules

In chapter 6 the theory of molecular orbitals and its application to several types of molecules was described in outline. One of the neatest illustrations of the use of molecular orbital theory is in the theory of *concerted reactions:* these are reactions which proceed in a single step in which all breakage and formation of bonds occur simultaneously. The Woodward-Hoffmann rules can be used to predict whether many concerted reactions can proceed *thermally*, that is with low activation energy which is insufficient to put any of the reacting molecules into an excited electronic state, or *photochemically*, that is with high activation energy and involving reacting molecules in excited electronic states. The basic premiss which underlies the rules is that *orbital symmetry is conserved* in these reactions. This means that for all elements of symmetry common to the reactant and product molecules the path of the reaction will be such that the symmetry of all the occupied orbitals with respect to these elements will be unchanged from reactants to products. From the consideration of electronic state symmetries in section 6.6 it is clear that if orbital symmetry is conserved then state symmetry must also be conserved. However, the converse is, in general, not true. For example, if we consider a single reactant forming a single product and preserving only a plane of symmetry then the orbital configuration change

$$(a')^2(a'')^2 \rightarrow (a'')^2(a')^2$$

would conserve orbital symmetry and also state symmetry since the states of reactant and product are both A′. But in the configuration change

$$(a')^2(a'')^2 \rightarrow (a')^2(a')^2$$

orbital symmetry is not conserved although state symmetry is, since again both states are A′.

A good example of the application of these principles is in the concerted cyclo-addition of two ethylene molecules to form cyclobutane, illustrated in figure 7.1. In order that the reaction can take place in a single step, the two ethylene molecules must approach each other in the symmetrical way illustrated in figure 7.1(a). This configuration as a whole belongs to the $\mathbf{D_{2h}}$

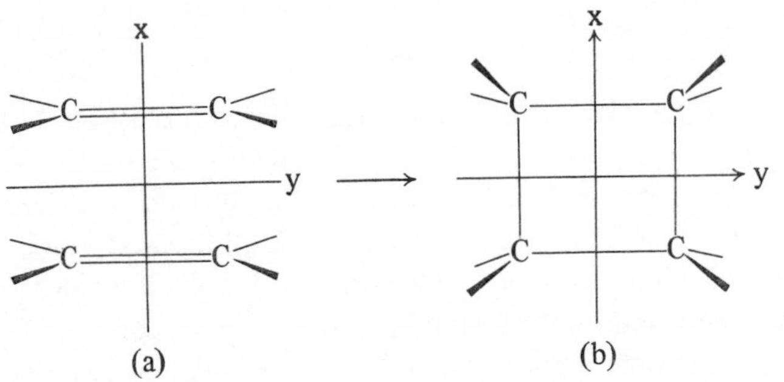

FIGURE 7.1
Axis notation for the concerted cycloaddition of two ethylene molecules (a) to form cyclobutane (b)

point group and, since it is non-planar, the labelling of axes is arbitary: that which has been adopted here is given in the figure. Cyclobutane belongs to the $\mathbf{D_{4h}}$ point group. (The ring is in fact slightly puckered, but this does not affect the labelling of orbitals according to the $\mathbf{D_{4h}}$ point group since the barrier (see section 7.8) to flattening the ring is low.) The elements of symmetry which are conserved in the reaction are *all* those of the $\mathbf{D_{2h}}$ point group, namely $\sigma(xy)$, $\sigma(xz)$, $\sigma(yz)$, i, $C_2(z)$, $C_2(y)$, $C_2(x)$, but in classifying the orbitals of reactants and product with respect to these elements we need only consider a set of generating elements of the $\mathbf{D_{2h}}$ point group, say the three planes of symmetry.

As a result of the formation of cyclobutane from two ethylene molecules the π orbitals of the latter have been replaced by σ orbitals, parallel to the x-axis, in the former. It is the correlation between reactants and products of these orbitals only which we need to consider since all the other orbitals are changed in a relatively minor way. The π orbitals, π and π^*, on each ethylene

		$\sigma(xy)$	$\sigma(xz)$	$\sigma(yz)$	Orbital classification
$\pi_1 + \pi_2$	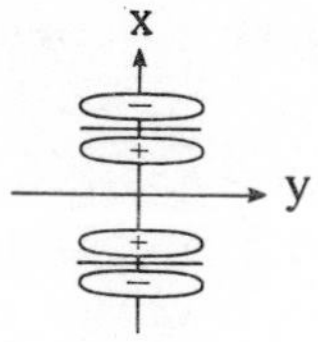	1	1	1	a_g(SS)
$\pi_1 - \pi_2$	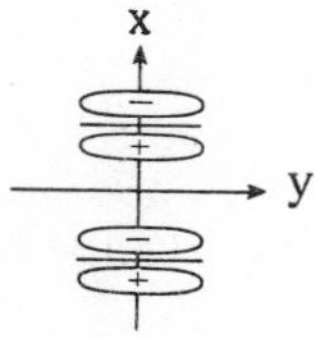	1	1	–1	b_{3u}(SA)
$\pi_1^* + \pi_2^*$	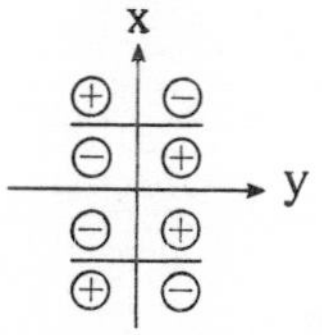	1	–1	1	b_{2u}(AS)
$\pi_1^* - \pi_2^*$	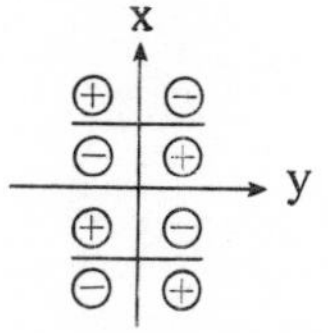	1	–1	–1	b_{1g}(AA)

FIGURE 7.2
Symmetry classification of π-MO's of two ethylene molecules in a $\mathbf{D_{2h}}$ configuration

molecule are illustrated in figure 6.8. For two molecules in the configuration illustrated in figure 7.1(a) the orbitals must be delocalized by taking in-phase and out-of-phase combinations of both pairs of π and π^* orbitals. These delocalized orbitals are illustrated and classified in figure 7.2. The orbital classifications are given both according to the symmetry species of the $\mathbf{D_{2h}}$ point group and also in a shorthand notation which does not require a detailed knowledge of point groups and which has commonly been adopted in discussions of the Woodward-Hoffmann rules. The notation recognizes that all the orbitals we shall be concerned with in this reaction are symmetric

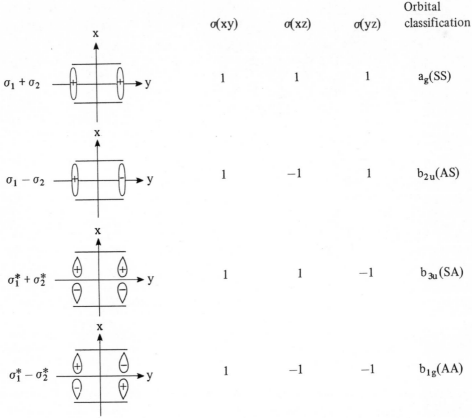

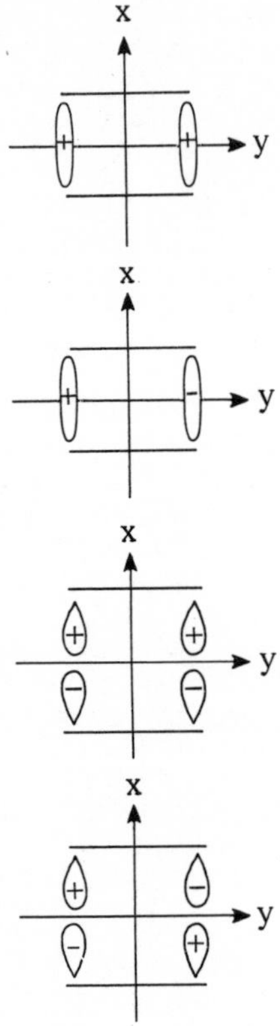

	$\sigma(xy)$	$\sigma(xz)$	$\sigma(yz)$	Orbital classification
$\sigma_1 + \sigma_2$	1	1	1	a_g(SS)
$\sigma_1 - \sigma_2$	1	−1	1	b_{2u}(AS)
$\sigma_1^* + \sigma_2^*$	1	1	−1	b_{3u}(SA)
$\sigma_1^* - \sigma_2^*$	1	−1	−1	b_{1g}(AA)

FIGURE 7.3
Symmetry classification of the σ-MO's which are formed when two ethylene molecules react to give cyclobutane

with respect to σ(xy) and S or A labels are used to imply respectively symmetry or antisymmetry with respect to the other two planes of symmetry.

Figure 7.3 illustrates the delocalized σ and σ* orbitals of cyclobutane which are formed in the reaction and gives their classifications.

In figure 7.4 the reactant and product orbitals are each arranged in order of increasing energy and correlated. The rules governing correlations are (a) that only orbitals of the same symmetry can be correlated, (b) that lines of

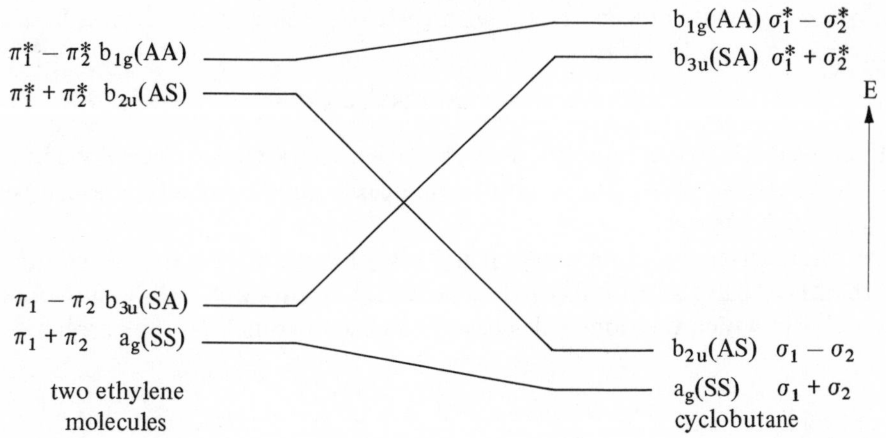

FIGURE 7.4
Correlation of MO's in the cycloaddition of two ethylene molecules to form cyclobutane

correlation corresponding to the same symmetry cannot cross (the *non-crossing rule*), and (c) that, provided (a) and (b) are obeyed, correlations are in order of increasing energy. It should be pointed out here that the energy scale in figure 7.4 is meant to give only a rough idea of the relative separations of orbitals: quantitative calculations need not concern us here.

Each of the orbitals of figure 7.4 can take two electrons so the ground state configuration of two ethylene molecules, is

$$(a_g)^2(b_{3u})^2 \text{ or } (SS)^2(SA)^2$$

and that of cyclobutane is

$$(a_g)^2(b_{2u})^2 \text{ or } (SS)^2(AS)^2$$

Clearly orbital symmetry is *not* conserved in a thermal conversion of ethylene into cyclobutane (or *vice versa*) and the reaction is said to be *symmetry for-*

bidden. On the other hand, if an electron is promoted in the ethylene molecules from the b_{3u} to the b_{2u} orbital, the configuration

$$(a_g)^2(b_{3u})(b_{2u}) \text{ or } (SS)^2(SA)(AS)$$

of this excited state does correlate with the first excited state of cyclobutane, which has the configuration

$$(a_g)^2(b_{2u})(b_{3u}) \text{ or } (SS)^2(AS)(SA)$$

with conservation of symmetry. Therefore the reaction can proceed via an excited electronic state, that is photochemically, and this reaction is said to be *symmetry allowed.*

An alternative way of demonstrating this is by correlating the electronic states rather than the orbitals of the reactants and product. This is shown in figure 7.5 in which the same rules have been used for making the correlations

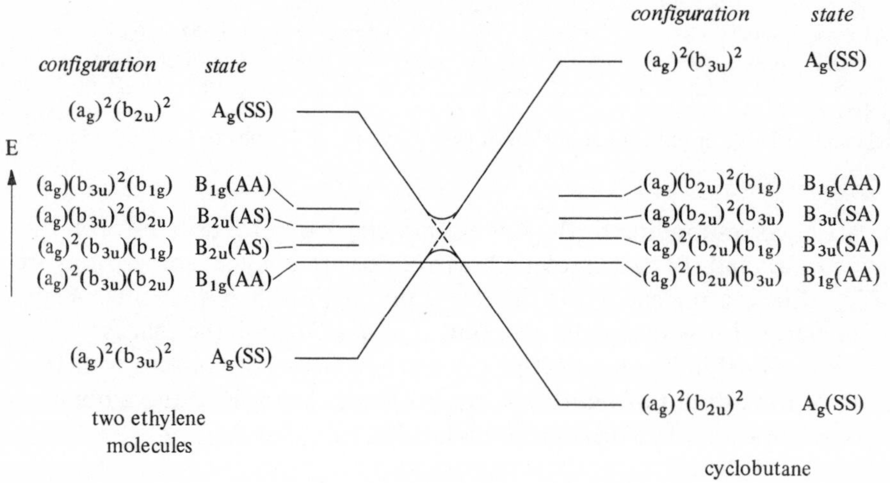

FIGURE 7.5
Correlation of electronic states in the cycloaddition of two ethylene molecules to form cyclobutane

as were used for correlating orbitals. Note the avoided crossing of the A_g state correlation lines as required by the non-crossing rule and that the state labels using A and S are obtained using the rules A x A = S, S x S = S, and

A × S = A. The energy barrier in going from the ground state of two ethylene molecules to that of cyclobutane is seen to be large while there is no barrier to the reaction via the first excited state.

Another type of reaction illustrating the use of the Woodward-Hoffmann rules is a concerted electrocyclic reaction. Such a reaction involves the formation of a single bond between the two ends of a non-branching conjugated π-electron system to give a cyclic molecule (or the reverse process in which the ring is opened). An example of an electrocyclic reaction is the conversion of cyclobutene to *s-cis*-butadiene illustrated in figure 7.6. In this

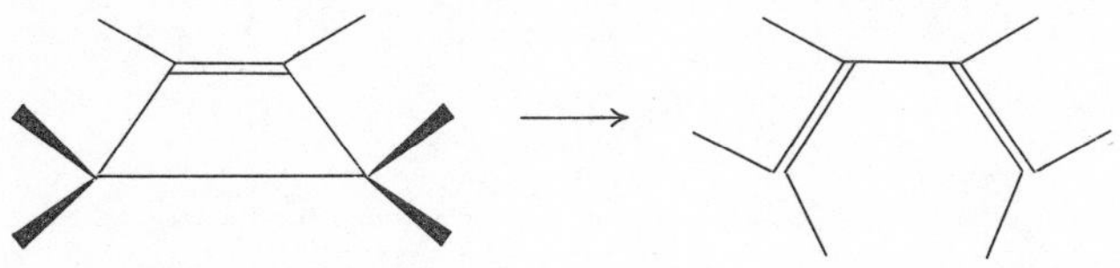

cyclobutene　　　　*s-cis*-butadiene

FIGURE 7.6
Isomerization of cyclobutene to *s-cis*-butadiene

reaction one double bond is lost and two are gained and the four out-of-plane hydrogen atoms of cyclobutene swing into the plane in *s-cis*-butadiene. As shown in figure 7.7 this process may occur in four ways of which two are called *conrotatory* (one clockwise, one anticlockwise) and the other two are called *disrotatory*. Unless the four atoms can be labelled in some way the products resulting from the four possible processes will be indistinguishable. In order to apply rigidly the method of orbital correlation to the problem of whether a conrotatory or disrotatory process will occur thermally or photochemically, we require to label the atoms in such a way that the electronic wave function is not perturbed. Isotopic substitution of deuterium for atom 1 and tritium for atom 3 will serve this purpose. This gives four distinguishable *s-cis*-butadiene molecules. However, consideration of orbital symmetry will enable us to decide between a con- or dis-rotatory process, but not between the two types of each. If substitution of other groups such as fluorine or methyl were made for the hydrogen atoms then stereochemical and electrostatic effects would probably decide between these two types.

Cyclobutene and *s-cis*-butadiene both belong to the $\mathbf{C}_{2v}$ point group.

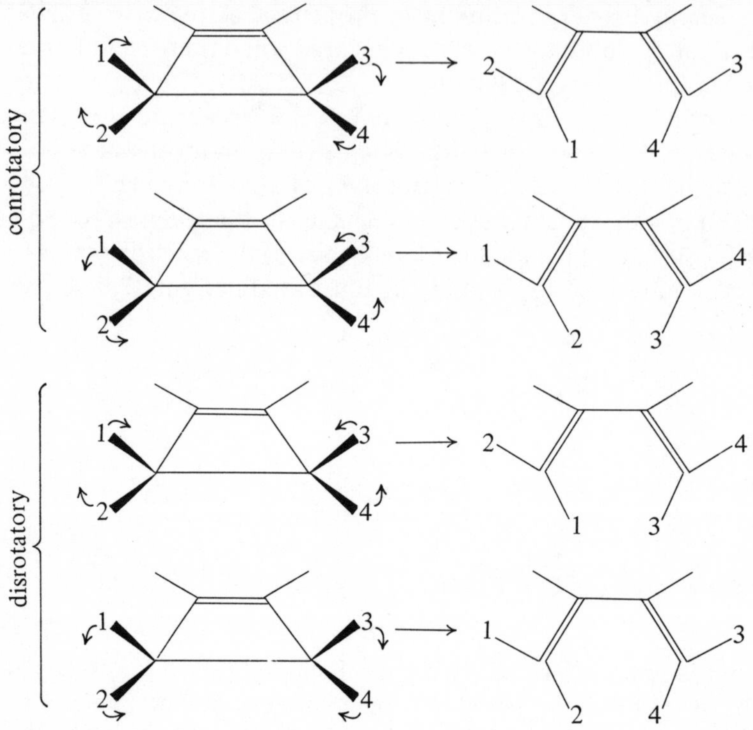

FIGURE 7.7
Conrotatory and disrotatory modes of isomerization of cyclobutene to *s-cis*-butadiene

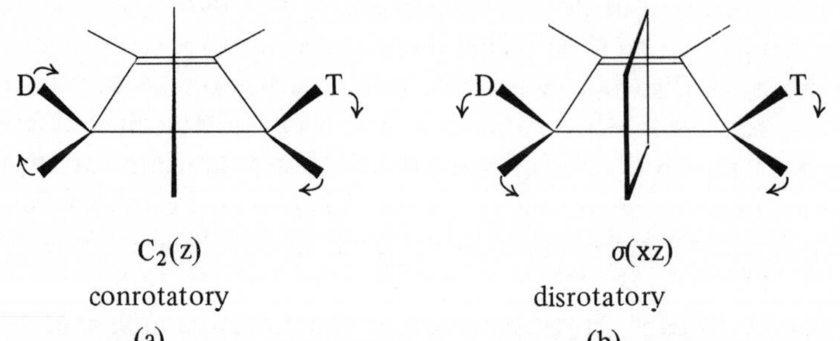

$C_2(z)$ conrotatory (a)

$\sigma(xz)$ disrotatory (b)

FIGURE 7.8
In the conrotatory mode the $C_2(z)$ axis is conserved and in the disrotatory mode the $\sigma(xz)$ plane is conserved

Throughout the conrotatory mode of reaction the $C_2(z)$ axis (figure 7.8(a)) is retained and therefore orbital symmetry with respect to this axis must be conserved. In the disrotatory mode it is the $\sigma(xz)$ plane (figure 7.8(b)) which is retained.

In the change from cyclobutene to *s-cis*-butadiene the most important orbitals in cyclobutene are the σ and σ^* orbitals in the broken σ bond and the π and π^* orbitals, which are very similar to those of ethylene, in the double bond. These orbitals are illustrated and classified according to the $\mathbf{C_2}$ point group (for the conrotatory mode) and the $\mathbf{C_s}$ point group (for the disrotatory mode) in figure 7.9. Once again the alternative (and commonly

		$C_2(z)$	$\sigma(xz)$	$\mathbf{C_2}$	$\mathbf{C_s}$
σ	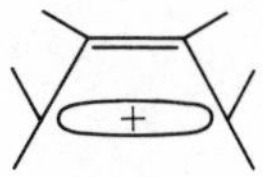	1	1	a(S)	a′(S)
π	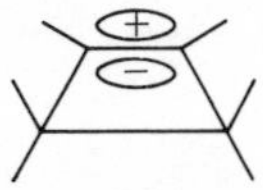	−1	1	b(A)	a′(S)
π^*	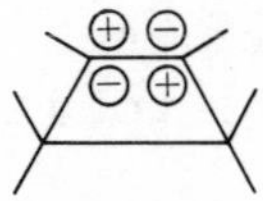	1	−1	a(S)	a″(A)
σ^*	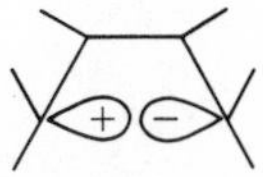	−1	−1	b(A)	a″(A)

FIGURE 7.9
Classification, according to the $\mathbf{C_2}$ and $\mathbf{C_s}$ point groups, of MO's of cyclobutene which are important in the isomerization reaction

used) S or A labels are given for the orbitals where S and A imply respectively symmetry or antisymmetry with respect to either the C_2 axis or σ plane.

In figure 7.10 the π orbitals of *s-cis*-butadiene, which have the form given in figure 6.12, are illustrated and they are classified according to the $\mathbf{C_2}$ and $\mathbf{C_s}$ point groups.

		$C_2(z)$	$\sigma(xz)$	$\mathbf{C_2}$	$\mathbf{C_s}$
ψ_1	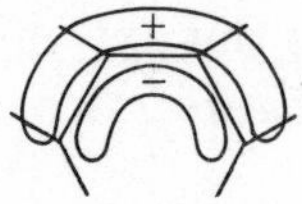	−1	1	b(A)	a′(S)
ψ_2	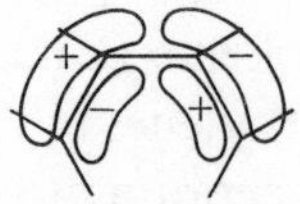	1	−1	a(S)	a″(A)
ψ_3	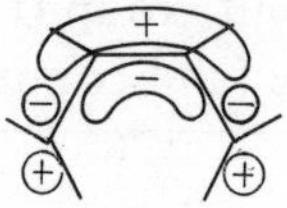	−1	1	b(A)	a′(S)
ψ_4	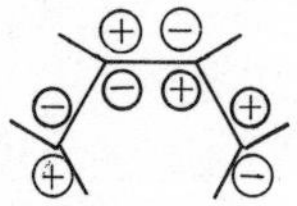	1	−1	a(S)	a″(A)

FIGURE 7.10
Classification of π-MO's of *s-cis*-butadiene according to the $\mathbf{C_2}$ and $\mathbf{C_s}$ point groups. (These orbitals are the same as those illustrated in figure 6.12)

In figures 7.11 and 7.12 the orbitals are correlated according to conrotatory and disrotatory modes respectively. Since there are four electrons to be accommodated in these orbitals the ground state configuration of cyclobutene, relevant to the conrotatory mode, is

$$(a)^2(b)^2 \text{ or } (S)^2(A)^2$$

and that of *s-cis*-butadiene is

$$(b)^2(a)^2 \text{ or } (A)^2(S)^2$$

The reaction, therefore, can proceed thermally in the conrotatory mode since orbital symmetry is conserved between both ground states. On the other hand the ground state configuration of cyclobutene, relevant to the disrotatory mode, is

$$(a')^2(a')^2 \text{ or } (S)^2(S)^2$$

whereas that of *s-cis*-butadiene is

$$(a')^2(a'')^2 \text{ or } (S)^2(A)^2$$

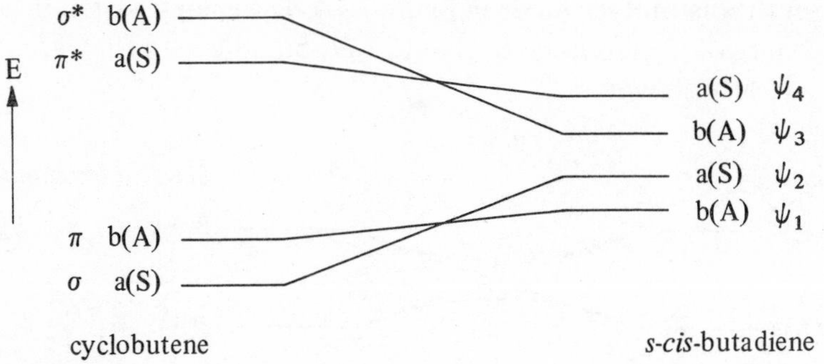

FIGURE 7.11
Correlation of MO's of cyclobutene and *s-cis*-butadiene in the conrotatory mode

The first excited configuration of cyclobutene relevant to the disrotatory mode is

$$(a')^2(a')(a'') \text{ or } (S)^2(S)(A)$$

and that of *s-cis*-butadiene is

$$(a')^2(a'')(a') \text{ or } (S)^2(A)(S)$$

Therefore the reaction can proceed photochemically but not thermally in the disrotatory mode.

States, rather than orbitals, are correlated for the conrotatory mode in

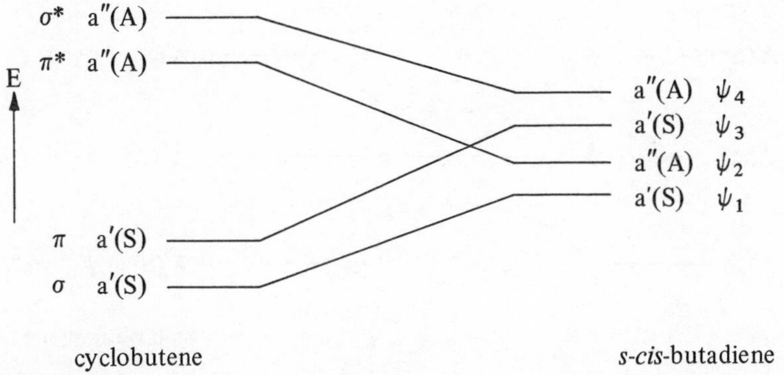

FIGURE 7.12
Correlation of MO's of cyclobutene and *s-cis*-butadiene in the disrotatory mode

figure 7.13 and for the disrotatory mode in figure 7.14. The energy barrier in going from the ground state of cyclobutene to that of *s-cis*-butadiene in the disrotatory mode is clearly shown.

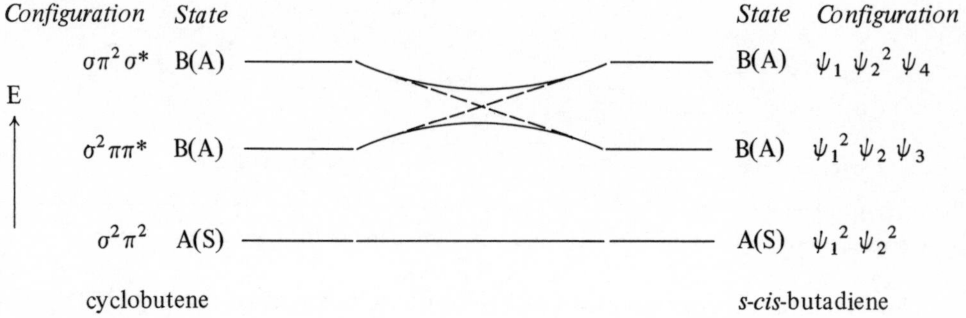

FIGURE 7.13
Correlation of electronic states of cyclobutene and *s-cis*-butadiene in the conrotatory mode

In many cases, though, the Woodward-Hoffmann rules have been applied to systems similar to those considered above but in which hydrogen atoms have been replaced by methyl, ethyl, fluorine, chlorine, bromine, etc. Indeed, in the conversion of cyclobutene to *s-cis*-butadiene it is only in

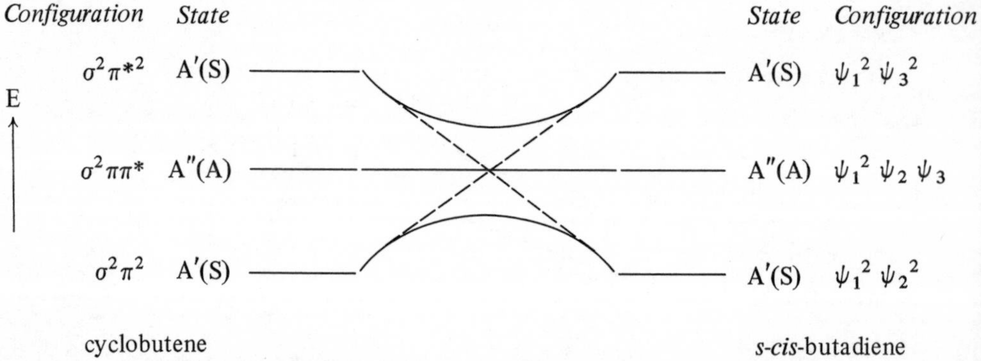

FIGURE 7.14
Correlation of electronic states of cyclobutene and *s-cis*-butadiene in the disrotatory mode

cases where hydrogen atoms have been replaced by substituents that it is important chemically whether the reaction proceeds by the con- or dis-rotatory mode. When such substitutions are made the rigid symmetry arguments used here break down because symmetry elements are lost. However, it appears that most substituents perturb the systems by an amount which is insufficient to invalidate energy level correlations of the type used in figures 7.4, 7.5, 7.11, 7.12, 7.13, and 7.14.

The reader is referred to 'The Conservation of Orbital Symmetry' by R. B. Woodward and R. Hoffmann (Verlag Chemie, Academic Press, 1970) for further examples of the application of the Woodward-Hoffmann rules to concerted reactions.

7.2 Electric dipole selection rules in molecules

In an absorption process either the electric or magnetic component of the electromagnetic radiation may interact with the atomic or molecular sample. In optical spectroscopy generally interaction with the electric component is of the order of 10^5 times stronger than with the magnetic component. Consequently transitions involving interaction with the electric component are observed much more commonly. Similarly, emission processes also usually involve the electric component of the radiation.

The interaction of the electric component of the radiation is with the electric dipole moment (or possibly the electric quadruple moment, but this type of interaction is usually extremely weak) of the atom or molecule. The selection rules governing transitions between states and involving such interaction are called *electric dipole selection rules.* We shall be deriving these for electronic, vibronic and vibrational transitions.

The transition moment **P** for an electronic, vibronic or vibrational transition between a lower state, described by the wave function ψ'' and an upper state described by the wave function ψ', is given by

$$\mathbf{P} = \int \psi'^* \mathbf{p} \psi'' \, d\tau \tag{7.1}$$

where **p** is the electric dipole moment (the symmetry classification of the components of **p** along the cartesian axes was discussed in section 5.2). Since the probability of a transition is proportional to $\mathbf{P}^2$, *for an allowed transition*

$\mathbf{P} \neq \mathbf{O}$ *and for a forbidden transition* $\mathbf{P} = \mathbf{O}$. On resolving the dipole moment into three components, equation 7.1 can be rewritten

$$P_x = \int \psi'^* p_x \psi'' \, d\tau$$
$$P_y = \int \psi'^* p_y \psi'' \, d\tau \qquad (7.2)$$
$$P_z = \int \psi'^* p_z \psi'' \, d\tau$$

from which it follows that a transition is allowed if either P_x, P_y, or P_z is non-zero. If only one component is non-zero then the transition is *polarized* along a cartesian axis, if two components are non-zero it is polarized in the xy-, yz- or xz-plane and if all three are non-zero it is not polarized in any restricted direction.

The requirement that an integral of the type in equations 7.1 and 7.2 is non-zero is that (a) *for transitions between non-degenerate states the symmetry species of the quantity inside the integral is totally symmetric* or (b) *for transitions between states of which at least one is degenerate the symmetry species of the quantity inside the integral should contain the totally symmetric symmetry species.* These rules can be written as

$$\text{(a)} \quad \Gamma(\psi') \times \Gamma(\mathbf{p}) \times \Gamma(\psi'') = \text{A} \qquad (7.3)$$

$$\text{(b)} \quad \Gamma(\psi') \times \Gamma(\mathbf{p}) \times \Gamma(\psi'') \supset \text{A} \qquad (7.4)$$

where Γ stands for 'symmetry species of . . .', A is used to denote any totally symmetric symmetry species and the Boolean symbol $\supset$ means 'contains'. It was shown in section 5.2 that

$$\Gamma(p_x) = \Gamma(\text{T}_x)$$
$$\Gamma(p_y) = \Gamma(\text{T}_y) \qquad (7.5)$$
$$\Gamma(p_z) = \Gamma(\text{T}_z)$$

7.2.1 ELECTRONIC TRANSITIONS BETWEEN NON-DEGENERATE STATES OF THE SAME MULTIPLICITY

If an electronic transition between two non-degenerate states of the same multiplicity ($2S + 1$ where S is the total electron spin quantum number) is to be allowed then, from equations 7.3 and 7.5,

$$\Gamma(\psi_e') \times \Gamma(T_x) \times \Gamma(\psi_e'') = A$$

and/or $$\Gamma(\psi_e') \times \Gamma(T_y) \times \Gamma(\psi_e'') = A \qquad (7.6)$$

and/or $$\Gamma(\psi_e') \times \Gamma(T_z) \times \Gamma(\psi_e'') = A$$

the 'and/or' implying that, for an allowed transition, one, two or three components of the transition moment **P** may be non-zero.

If we consider initially only electronic transitions in which the lower state is the totally symmetric ground state then $\Gamma(\psi_e'') = A$ and equation 7.6 becomes

$$\Gamma(\psi_e') \times \Gamma(T_x) = A$$

and/or $$\Gamma(\psi_e') \times \Gamma(T_y) = A \qquad (7.7)$$

and/or $$\Gamma(\psi_e') \times \Gamma(T_z) = A$$

from which it follows that

$$\Gamma(\psi_e') = \Gamma(T_x) \text{ and/or } \Gamma(T_y), \text{ and/or } \Gamma(T_z) \qquad (7.8)$$

This equation represents a very simple rule by which the allowed electronic transitions from the totally symmetric ground state of any molecule can be obtained by looking up $\Gamma(T_x)$, etc. in the appropriate character table. For example, for a molecule belonging to the $\mathbf{C_{2v}}$ point group table 4.11 gives the $\Gamma(T_x)$, etc. and the allowed transitions are

$$B_1 - A_1$$

$$B_2 - A_1 \qquad (7.9)$$

$$A_1 - A_1$$

(It should be noted that in denoting transitions between two states the upper state is written first and the lower state second. However, if the transition P–Q is allowed then so is the transition Q–P). The $B_1 - A_1$ transition is polarized along the x-axis, $B_2 - A_1$ along the y-axis and $A_1 - A_1$ along the z-axis. Similarly, in the $\mathbf{D_{2h}}$ point group the transitions

$$\begin{aligned} B_{3u} - A_g \\ B_{2u} - A_g \\ B_{1u} - A_g \end{aligned} \tag{7.10}$$

are allowed and polarized along the x-, y-, and z-axes respectively. In the $\mathbf{C_s}$ point group the transitions

$$\begin{aligned} A' - A' \\ A'' - A' \end{aligned} \tag{7.11}$$

are allowed and polarized respectively in the xy-plane (since $\Gamma(T_x) = \Gamma(T_y) = A'$) and along the z-axis.

If the lower state is not totally symmetric this means in most cases that it is not the ground state and selection rules will be concerned with transitions between different excited states. For such transitions to be allowed it follows from equation 7.6 that

$$\Gamma(\psi_e') \times \Gamma(\psi_e'') = \Gamma(T_x), \text{ and/or } \Gamma(T_y), \text{ and/or } \Gamma(T_z) \tag{7.12}$$

For example if, for a molecule in the $\mathbf{C_{2v}}$ point group, $\Gamma(\psi_e'') = A_2$ the allowed transitions are, from equation 7.12 and using the rules in section 4.3.1 for the multiplication of symmetry species,

$$\begin{aligned} B_2 - A_2 \\ B_1 - A_2 \\ A_2 - A_2 \end{aligned} \tag{7.13}$$

These transitions are polarized along the x-, y-, and z-axes respectively.

For a molecule in the $\mathbf{D_{2h}}$ point group, if $\Gamma(\psi_e'') = B_{1g}$, the allowed transitions are

$$\begin{aligned} B_{2u} - B_{1g} \\ B_{3u} - B_{1g} \\ A_u - B_{1g} \end{aligned} \tag{7.14}$$

polarised along the x-, y-, and z-axes respectively.

7.2.2 ELECTRONIC TRANSITIONS BETWEEN STATES OF THE SAME MULTIPLICITY AND OF WHICH AT LEAST ONE IS DEGENERATE

If a transition of this type is to be allowed then, from equations 7.4 and 7.5,

$$\Gamma(\psi'_e) \times \Gamma(T_x) \times \Gamma(\psi''_e) \supset A$$

and/or $$\Gamma(\psi'_e) \times \Gamma(T_y) \times \Gamma(\psi''_e) \supset A \qquad (7.15)$$

and/or $$\Gamma(\psi'_e) \times \Gamma(T_z) \times \Gamma(\psi''_e) \supset A$$

If the lower state is the totally symmetric ground state, $\Gamma(\psi''_e) = A$ and equation 7.15 becomes

$$\Gamma(\psi'_e) \times \Gamma(T_x) \supset A$$

and/or $$\Gamma(\psi'_e) \times \Gamma(T_y) \supset A \qquad (7.16)$$

and/or $$\Gamma(\psi'_e) \times \Gamma(T_z) \supset A$$

For example, in the $\mathbf{D}_{6h}$ point group, $\Gamma(T_x, T_y) = E_{1u}$. Table 4.52 shows that the only symmetry species which, when multiplied by E_{1u} contains A_{1g} is E_{1u}.† Since $\Gamma(T_z) = A_{2u}$ the allowed transitions involving an A_{1g} lower state are

$$E_{1u} - A_{1g}$$
$$A_{2u} - A_{1g} \qquad (7.17)$$

which are polarized respectively in the xy-plane and along the z-axis. *In fact in all degenerate point groups, as in non-degenerate point groups, all transitions are allowed between a totally symmetric state and a second state if the symmetry species of the second state is that of a translation.*

In the $\mathbf{O}_h$ point group

$$T_{1u} - A_{1g} \qquad (7.18)$$

is the only type of transition which is allowed involving a totally symmetric lower state: this polarization is not restricted since $\Gamma(T_x, T_y, T_z) = T_{1u}$.

In the $\mathbf{D}_{\infty h}$ point group

$$\Pi_u - \Sigma_g^+$$
$$\Sigma_u^+ - \Sigma_g^+ \qquad (7.19)$$

† It is a useful general rule that if $\Gamma_i \times \Gamma_j \supset A$ then $\Gamma_i = \Gamma_j$.

are the allowed transitions involving a Σ_g^+ lower state. They are polarized respectively in the xy-plane and along the z-axis. These selection rules are consistent with the quantum number selection rule $\Delta\Lambda = 0, \pm 1$ where $\Lambda = 0, 1, 2, 3, \ldots$ in $\Sigma, \Pi, \Delta, \Phi \ldots$ states respectively and Λ is a quantum number associated with the orbital angular momentum of the electrons.

Transitions which do not involve a totally symmetric state will usually be between excited states. It follows from equation 7.15 that, for such transitions to be allowed,

$$\Gamma(\psi_e') \times \Gamma(\psi_e'') \supset \Gamma(T_x), \text{ and/or } \Gamma(T_y), \text{ and/or } \Gamma(T_z), \tag{7.20}$$

For example, in the $\mathbf{D}_{6h}$ point group, allowed transitions involving an A_{2g} state are

$$\begin{array}{c} E_{1u} - A_{2g} \\ A_{1u} - A_{2g} \end{array} \tag{7.21}$$

polarized respectively in the xy-plane and along the z-axis.† If the lower state symmetry species is E_{2u} then, using equation 7.20 and table 4.52, the allowed transitions are

$$\begin{array}{c} B_{1g} - E_{2u} \\ B_{2g} - E_{2u} \\ E_{1g} - E_{2u} \\ E_{2g} - E_{2u} \end{array} \tag{7.22}$$

the first three of which are polarized in the xy-plane and the fourth along the z-axis.

In the $\mathbf{C}_{3v}$ point group, transitions from an E state to all other states are allowed

$$\begin{array}{c} A_1 - E \\ A_2 - E \\ E - E \end{array} \tag{7.23}$$

† It is a useful rule that if $\Gamma_i \times \Gamma_j \supset \Gamma(T)$ then $\Gamma_i = \Gamma(T) \times \Gamma_j$ and $\Gamma_j = \Gamma(T) \times \Gamma_i$.

Both the $A_1 - E$ and $A_2 - E$ transitions are polarized in the xy-plane but the $E - E$ transition has two allowed components since, from table 4.52,

$$E \times E \supset \Gamma(T_z) + \Gamma(T_x, T_y) \tag{7.24}$$

where $\Gamma(T_z) = A_1$ and $\Gamma(T_x, T_y) = E$.

In the $C_{\infty v}$ point group the allowed transitions from a Π lower state are

$$\begin{aligned} \Sigma^+ &- \Pi \\ \Sigma^- &- \Pi \\ \Delta &- \Pi \\ \Pi &- \Pi \end{aligned} \tag{7.25}$$

These are in agreement with the $\Delta\Lambda = 0, \pm 1$ selection rule. The $\Pi - \Pi$ transition is polarized along the z-axis and the other three in the xy-plane.

As a final example, in the O_h point group the following transitions from a T_{1g} state are allowed since all the products of the symmetry species of the combining states contain the species T_{1u} where $T_{1u} = \Gamma(T_x, T_y, T_z)$

$$\begin{aligned} A_{1u} &- T_{1g} \\ E_u &- T_{1g} \\ T_{1u} &- T_{1g} \\ T_{2u} &- T_{1g} \end{aligned} \tag{7.26}$$

All the transition polarizations are not restricted.

7.2.3 ELECTRONIC TRANSITIONS BETWEEN STATES OF DIFFERENT MULTIPLICITY

The selection rule $\Delta S = 0$ is not rigorously obeyed in molecules with atoms which are not of low atomic number. Many electronic transitions with $\Delta S = \pm 1$ in which the electron spin multiplicity $(2S + 1)$ changes by 2 have been observed. In such cases the electron is not only moving from one orbital to another but is also having the sense of its spin reversed.

Known transitions with $\Delta S = \pm 1$ are mainly between singlet and triplet states, for example the ${}^3A_2 - {}^1A_1$ transition in formaldehyde (see section 6.6). There are also a few known transitions between doublet and quartet states, for example the $^4\Sigma^- - {}^2\Pi$ transition in SiF, but they are uncommon and we shall not consider them further.

Transitions with $\Delta S = \pm 1$ are very weak compared to those with $\Delta S = 0$ except in molecules where a heavy atom such as iodine is present. Transitions with $\Delta S = \pm 2$, for example between singlet and quintet states, are weaker still and will not be discussed at all here.

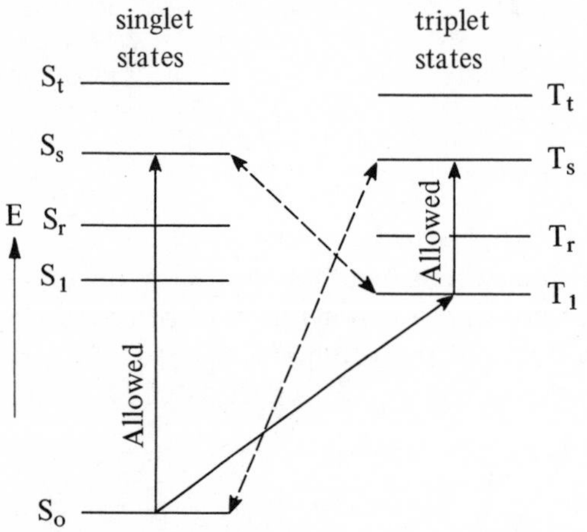

FIGURE 7.15
Effect of spin-orbit coupling on a transition between the lowest singlet (S_o) and lowest triplet (T_1) states. Mixing of states is indicated by broken lines

The $\Delta S = 0$ selection rule is broken down by spin-orbit coupling, the coupling between the spin and orbital motion of the electrons. Spin-orbit coupling always occurs to a greater or lesser degree and, when it is taken into account, states which were pure singlet or triplet states in the approximation of zero coupling become mixed. The extent to which the $\Delta S = 0$ selection rule breaks down depends on the degree of coupling. The most commonly observed type of transition between a triplet and a singlet state is between the singlet ground state S_0 and the first excited triplet state T_1: this is illustrated in figure 7.15. Spin-orbit coupling mixes states in the manifold of singlet states with those in the triplet manifold but, for a particular state, mixing will usually occur preferentially with only one other state because of restrictions of symmetry and energy separation, which must not be too large. The mixing results in the intensity of an otherwise spin-forbidden triplet-singlet transition being 'stolen' from an allowed transition from the state with which mixing occurs. In the case of the $T_1 - S_0$ transition in figure 7.15, state S_0 may mix preferentially with one triplet state, T_s, to which there is an allowed transition from T_1. Intensity in the $T_1 - S_0$ transition is then stolen from the transition $T_s - T_1$. Alternatively, or in addition, state T_1 may mix preferentially with one singlet state S_s, to which there is an allowed transition from S_0. In this case, intensity in the $T_1 - S_0$ transition is stolen from the transition $S_s - S_0$. The various mixing processes are shown by broken lines in figure 7.15.

It should be pointed out that the stealing of intensity from one transition by another represents a completely artificial description of the situation. The artificiality results from the initial approximation of zero spin-orbit coupling. It is only because of the general usefulness of this approximation that the concept of intensity stealing arises.

The appearance of a triplet-singlet transition with zero or non-zero intensity is determined by the absence or presence of mixing of states by spin-orbit coupling of the type illustrated in figure 7.15. Whether spin-orbit coupling can or cannot mix states is determined completely by symmetry arguments. The symmetry of a triplet state is the product of the spin and orbital symmetry. Each of the three spin components has the symmetry species of a rotation so that *a triplet-singlet transition might be observed if either the triplet state is connected by a rotational symmetry species to a nearby singlet state or the singlet state is similarly connected to a nearby triplet state.* This rule can be applied to the $T_1 - S_0$ transition in figure 7.15 by saying that the transition may have non-zero intensity if

$$\begin{aligned}
&& \Gamma(S_0) \times \Gamma(R_x) &= \Gamma(T_s) \\
&\text{and/or} & \Gamma(S_0) \times \Gamma(R_y) &= \Gamma(T_s) \\
&\text{and/or} & \Gamma(S_0) \times \Gamma(R_z) &= \Gamma(T_s) \\
&& \text{-----------------} & \\
&\text{and/or} & \Gamma(S_s) \times \Gamma(R_x) &= \Gamma(T_1) \\
&\text{and/or} & \Gamma(S_s) \times \Gamma(R_y) &= \Gamma(T_1) \\
&\text{and/or} & \Gamma(S_s) \times \Gamma(R_z) &= \Gamma(T_1)
\end{aligned} \qquad (7.27)$$

As an example we will consider the transition from the ground state of formaldehyde to the first triplet excited state, ${}^3A_2 - {}^1A_1$. Formaldehyde belongs to the C_{2v} point group (table 4.11) and in this case $\Gamma(S_0) = A_1$, $\Gamma(T_1) = A_2$, $\Gamma(R_x) = B_2$, $\Gamma(R_y) = B_1$, $\Gamma(R_z) = A_2$ and equations 7.27 give respectively

$$\begin{array}{c} \Gamma(S) \times \Gamma(R) = \Gamma(T) \\ \hline A_1 \times B_2 = B_2 \\ A_1 \times B_1 = B_1 \\ A_1 \times A_2 = A_2 \\ \hdashline B_1 \times B_2 = A_2 \\ B_2 \times B_1 = A_2 \\ A_1 \times A_2 = A_2 \end{array} \tag{7.28}$$

Therefore the ${}^3A_2 - {}^1A_1$ transition may steal intensity by any of the six mechanisms of equation 7.28 i.e. from any of the six types of transitions given in table 7.1, all of which are allowed, as can be shown by the rules

TABLE 7.1

Transition	*Polarization*
${}^3B_2 - {}^3A_2$	x-polarized
${}^3B_1 - {}^3A_2$	y-polarized
${}^3A_2 - {}^3A_2$	z-polarized
${}^1B_1 - {}^1A_1$	x-polarized
${}^1B_2 - {}^1A_1$	y-polarized
${}^1A_1 - {}^1A_1$	z-polarized

discussed in section 7.2.1. In the case of formaldehyde there is one mechanism which is much more effective in intensity stealing than the other five. This is the last of the six in table 7.1 in which intensity is stolen from a ${}^1A_1 - {}^1A_1$ transition. It is known that the second excited singlet state of formaldehyde is a ${}^1A_1$ state: this is quite close to the ${}^3A_2$ state and relatively strong mixing occurs by spin-orbit coupling. For this reason it is observed that the ${}^3A_2 - {}^1A_1$ transition moment is polarized along the z-axis, which is the direction of polarization of the ${}^1A_1 - {}^1A_1$ transition. *It is an important general rule that the polarization of a spin-forbidden transition is the same as that of the transition(s) from which the intensity is stolen by spin-orbit coupling.* This may lead, in principle, to mixed polarization but it is often the case, as in formaldehyde, that one of the possible mixing mechanisms is dominant.

We take as a second example of a triplet-singlet transition in a molecule belonging to a non-degenerate point group a ${}^3A_u - {}^1A_g$ transition in a molecule belonging to the $\mathbf{D}_{2h}$ point group (see table 4.32 for character table). In this case equations 7.27 become

$$\begin{array}{c} \Gamma(S) \times \Gamma(R) = \Gamma(T) \\ \hline A_g \times B_{3g} = B_{3g} \\ A_g \times B_{2g} = B_{2g} \\ A_g \times B_{1g} = B_{1g} \\ \hdashline B_{3u} \times B_{3g} = A_u \\ B_{2u} \times B_{2g} = A_u \\ B_{1u} \times B_{1g} = A_u \end{array} \qquad (7.29)$$

The ${}^3A_u - {}^1A_g$ transition may steal intensity by the six mechanisms of equation 7.29 from the six types of symmetry-allowed transitions given in table 7.2.

TABLE 7.2

Transition	*Polarization*
${}^3B_{3g} - {}^3A_u$	x-polarized
${}^3B_{2g} - {}^3A_u$	y-polarized
${}^3B_{1g} - {}^3A_u$	z-polarized
${}^1B_{3u} - {}^1A_g$	x-polarized
${}^1B_{2u} - {}^1A_g$	y-polarized
${}^1B_{1u} - {}^1A_g$	z-polarized

The polarization of a $^3A_u - {}^1A_g$ transition will depend on which of the mechanisms is dominant.

In both the $^3A_2 - {}^1A_1$ and $^3A_u - {}^1A_g$ transitions discussed, spin-orbit coupling makes the transitions allowed whereas the corresponding singlet-singlet transitions $^1A_2 - {}^1A_1$ and $^1A_u - {}^1A_g$ are forbidden by electric dipole selection rules.

As an example of a triplet-singlet transition involving a degenerate state we consider a $^3E' - {}^1A_1'$ transition in a molecule belonging to the $\mathbf{D}_{3h}$ point group (character table given in table 4.33). Equations 7.27 become

$$
\begin{array}{c}
\Gamma(S) \times \Gamma(R) = \Gamma(T) \\
\hline
A_1' \times E'' = E'' \\
A_1' \times A_2' = A_2' \\
\hdashline
\left.\begin{array}{c} A_1'' \\ A_2'' \end{array}\right\} \times E'' = E' \\
E' \times A_2' = E'
\end{array}
\tag{7.30}
$$

The $^3E' - {}^1A_1'$ transition may steal intensity by the mechanisms of equations 7.30 from any of the transitions given in table 7.3. The rules discussed in sections 7.2.1 and 7.2.2 can be used to determine the allowed or forbidden character of the transitions in this table.

Quartet-doublet, quintet-triplet and other transitions in which $\Delta S = \pm 1$

TABLE 7.3

Transition	*Polarization*
$^3E'' - {}^3E'$	One component allowed (z-polarized), two components forbidden
$^3A_2' - {}^3E'$	xy-polarized
$^1A_1'' - {}^1A_1'$	forbidden
$^1A_2'' - {}^1A_1'$	z-polarized
$^1E' - {}^1A_1'$	xy-polarized

can be treated in a similar way to triplet-singlet transitions since the spin-orbit interaction again has the species of a rotation. This is not the case when $|\Delta S| > 1$ but examples of such transitions are very rare.

7.2.4 TRANSITIONS BETWEEN VIBRONIC STATES

For allowed vibronic transitions, involving electronic states in either or both of which there may be vibrational excitation, a slightly modified form of equation 7.6 holds for non-degenerate vibronic states, namely

$$\Gamma(\psi'_{ev}) \times \Gamma(T_x) \times \Gamma(\psi''_{ev}) = A$$

and/or $$\Gamma(\psi'_{ev}) \times \Gamma(T_y) \times \Gamma(\psi''_{ev}) = A \quad (7.31)$$

and/or $$\Gamma(\psi'_{ev}) \times \Gamma(T_z) \times \Gamma(\psi''_{ev}) = A$$

where the ψ_{ev} are vibronic wave functions. From these equations it follows that

$$\Gamma(\psi'_{ev}) \times \Gamma(\psi''_{ev}) = \Gamma(T_x) \text{ and/or } \Gamma(T_y) \text{ and/or } \Gamma(T_z) \quad (7.32)$$

which is analogous to equation 7.12. Similarly, if either or both of the vibronic states are degenerate then equation 7.32 is modified to

$$\Gamma(\psi'_{ev}) \times \Gamma(\psi''_{ev}) \supset \Gamma(T_x) \text{ and/or } \Gamma(T_y) \text{ and/or } \Gamma(T_z) \quad (7.33)$$

for an allowed transition.

If we assume that the Born-Oppenheimer approximation (section 1.2) holds the vibronic wave functions can be factorized into an electronic and a vibrational wave function

$$\psi_{ev} = \psi_e \times \psi_v \quad (7.34)$$

Then, for non-degenerate vibronic states, equation 7.32 becomes

$$\Gamma(\psi'_e) \times \Gamma(\psi'_v) \times \Gamma(\psi''_e) \times \Gamma(\psi''_v) = \Gamma(T_x) \text{ and/or } \Gamma(T_y) \text{ and/or } \Gamma(T_z) \quad (7.35)$$

If we consider an example of a vibronic transition in a molecule belonging to the $\mathbf{D}_{2h}$ point group in which $\Gamma(\psi_e') = B_{1g}$, $\Gamma(\psi'_v) = A_u$, $\Gamma(\psi''_e) = B_{3g}$, and $\Gamma(\psi''_v) = A_g$ then

$$\Gamma(\psi'_e) \times \Gamma(\psi'_v) \times \Gamma(\psi''_e) \times \Gamma(\psi''_v) = B_{2u} = \Gamma(T_y) \quad (7.36)$$

and we see that the transition is allowed. Note, however, that the pure electronic transition $B_{1g} - B_{3g}$ is forbidden.

For vibronic transitions involving degenerate states, applying the Born-Oppenheimer approximation to equation 7.33 gives

$$\Gamma(\psi'_e) \times \Gamma(\psi'_v) \times \Gamma(\psi''_e) \times \Gamma(\psi''_v) \supset \Gamma(T_x) \text{ and/or } \Gamma(T_y) \text{ and/or } \Gamma(T_z) \quad (7.37)$$

If, for example, in a molecule belonging to $\mathbf{D}_{6h}$ point group, $\Gamma(\psi'_e) = B_{2u}$, $\Gamma(\psi'_v) = E_{1g}$, $\Gamma(\psi''_e) = A_{1g}$, and $\Gamma(\psi''_v) = E_{2g}$ then

$$\Gamma(\psi'_e) \times \Gamma(\psi'_v) \times \Gamma(\psi''_e) \times \Gamma(\psi''_v) = A_{1u} + A_{2u} + E_{2u} \quad (7.38)$$

This product contains $\Gamma(T_z)$ which is A_{2u} and therefore the vibronic transition is allowed. However, in this case also, the pure electronic transition $B_{2u} - A_{1g}$ is forbidden.

7.2.5 TRANSITIONS BETWEEN VIBRATIONAL STATES

For pure vibrational transitions to be allowed between non-degenerate vibrational states then

$$\Gamma(\psi'_v) \times \Gamma(\psi''_v) = \Gamma(T_x) \text{ and/or } \Gamma(T_y) \text{ and/or } \Gamma(T_z) \quad (7.39)$$

which is analogous to equation 7.32. For example, if we consider in a molecule belonging to the $\mathbf{C}_{2v}$ point group a transition between a lower state, in which it is vibrating with one quantum of an a_2 vibration, to an upper state in which it is vibrating with one quantum each of a b_1 and a b_2 vibration then $\Gamma(\psi'_v) = b_1 \times b_2 = A_2$, and $\Gamma(\psi''_v) = A_2$. It follows that $\Gamma(\psi'_v) \times \Gamma(\psi''_v) = A_1$ and the transition is allowed and polarized along the z-axis.

For allowed vibrational transitions involving at least one degenerate state

$$\Gamma(\psi'_v) \times \Gamma(\psi''_v) \supset \Gamma(T_x) \text{ and/or } \Gamma(T_y) \text{ and/or } \Gamma(T_z) \quad (7.40)$$

which is analogous to equation 7.33. For example, a transition between a lower state in which a molecule belonging to the $\mathbf{C}_{3v}$ point group is vibrating with one quantum of an a_2 vibration and an upper state in which it is vibrating with two quanta of an e vibration then $\Gamma(\psi'_v) = (e)^2 = A_1 + E$, and $\Gamma(\psi''_v) = A_2$. It follows that $\Gamma(\psi'_v) \times \Gamma(\psi''_v) = A_2 + E$ which contains E which is $\Gamma(T_x, T_y)$ and the transition is allowed.

It is true, of course, that the general vibronic selection rules expressed by equations 7.32 and 7.33 apply to pure vibrational transitions and to pure electronic transitions between states of the same multiplicity: it is just that in pure electronic transitions the $\Gamma(\psi_v)$ can be ignored and in pure vibrational transitions the $\Gamma(\psi_e)$ can be ignored.

7.3 Electric dipole selection rules in atoms

Selection rules governing electronic transitions between singlet states in atoms are usually discussed in terms of the quantum number associated with the orbital angular momentum of the electrons: this quantum number is l for one electron and L for many electrons. For an atom or ion with one electron only e.g. H, He^+, Li^{2+}, the selection rule for electronic transitions is

$$\Delta l = \pm 1 \tag{7.41}$$

Since $l = 0, 1, 2, 3, 4 \ldots$ for s, p, d, f, g ... orbitals respectively the electron can transfer, for example, from a d orbital to a p or f orbital only. The allowed transitions are then

$$\begin{matrix} \mathrm{P - D} \\ \mathrm{F - D} \end{matrix} \tag{7.42}$$

In atoms such as sodium, in which there is one outer valence electron whose angular momentum is not strongly coupled to those of the other electrons, the selection rule is again $\Delta l = \pm 1$ for transitions which involve only the valence electron.

For polyelectronic atoms in general all the orbital angular momenta are coupled and the selection rule is

$$\Delta L = 0, \pm 1, \quad \text{except} \quad L = 0 \nleftrightarrow L = 0 \tag{7.43}$$

where L is the quantum number associated with the total electronic orbital angular momentum. There is an additional requirement that, if only one electron transfers from one orbital to another in the transition then, for this electron, $\Delta l = \pm 1$ only.

It is important to realize that electronic selection rules in atoms can also be discussed in terms of symmetry properties using the $\mathbf{K_h}$ point group whose character table is given in table 4.45. Indeed discussion of the selection rules in these terms is in many ways more satisfactory since it removes what appears otherwise to be a rather artificial division in discussing selection rules in atoms entirely in terms of quantum numbers and in polyatomic molecules in terms of symmetry properties.

Atomic orbitals commonly are labelled s, p, d, f, g . . . but if they are given the correct symmetry species according to the $\mathbf{K_h}$ point group they should be labelled s_g, p_u, d_g, f_u, g_g, . . . (the g or u character is obvious from the form of the orbitals).

$\mathbf{K_h}$ is a degenerate point group and selection rules can be derived using equation 7.20 provided that we know the rules for multiplying degenerate symmetry species in this point group. Multiplication of two species corresponding to L (or l) values of L_1 and L_2 (or l_1 and l_2) gives species with values $L_1 + L_2, L_1 + L_2 - 1, \ldots |L_1 - L_2|$. For example, if $L_1 = 1$ (symmetry species P_u) and $L_2 = 2$ (symmetry species D_g) then

$$P_u \times D_g = F_u + D_u + P_u \tag{7.44}$$

In equation 7.44 account has been taken of the usual rule that g x u = u. Similarly

$$F_u \times D_g = H_u + G_u + F_u + D_u + P_u \tag{7.45}$$

Since $\Gamma(T_x, T_y, T_z) = P_u$ equation 7.20 is satisfied for both transitions of equation 7.42 and they are allowed by symmetry. On the other hand, a promotion from a d-orbital to another d-orbital is forbidden since

$$D_g \times D_g = G_g + F_g + D_g + P_g + S_g \tag{7.46}$$

and this product does not contain a translational symmetry species. In this way the $\Delta l = \pm 1$ selection rule can be obtained by consideration of symmetry properties alone.

In polyelectronic atoms the possibility that $\Delta L = 0$ as well as ± 1 arises because, from suitable orbital configurations, S, P, D, F, G, . . . states which are *either* g or u states are possible. For example the states which arise from two electrons in d_g orbitals having different values of the principle quantum number n are given by

$$d_g \times d_g = G_g + F_g + D_g + P_g + S_g \tag{7.47}$$

The states arising from two electrons, one in a p_u orbital and one in a d_g orbital are given by

$$p_u \times d_g = F_u + D_u + P_u \tag{7.48}$$

Now a transition, for which $\Delta L = 0$, between the P_g state of equation (7.47) and the P_u state of equation 7.48 is allowed because

$$P_g \times P_u = D_u + P_u + S_u \tag{7.49}$$

which contains the translational species P_u. Similarly a $\Delta L = 0$ transition between the D_g state of equation 7.47 and the D_u state of equation 7.48 is allowed because

$$D_g \times D_u = G_u + F_u + D_u + P_u + S_u \qquad (7.50)$$

However S – S transitions will always be forbidden because $S \times S = S$ which cannot contain P_u.

The further qualification to the selection rules in equation 7.43, that if only one electron is transferred from one orbital to another then $\Delta l = \pm 1$ for this electron, follows also from the g – u selection rule. For example, if an electronic configuration changes from $3p_u 3d_g$ to $3p_u 4d_g$ then all the states arising from these two configurations are u states and u – u transitions are forbidden.

Summarizing, it can be said that electronic selection rules for *all* atoms, whether they have one or more electrons, can be expressed simply by saying that for transitions to be allowed between singlet states

$$\Gamma(\psi_e') \times \Gamma(\psi_e'') \supset P_u \qquad (7.51)$$

and the g – u selection rule must be obeyed.

7.4 Magnetic dipole selection rules

As was mentioned in section 7.2, in optical spectroscopy interaction of atoms or molecules with the electric component of the electromagnetic radiation is of the order of 10^5 times stronger than with the magnetic component. Therefore transitions which are allowed by magnetic dipole selection rules are very much weaker than those which are allowed by electric dipole selection rules. There are, however, several examples known in electronic spectra of diatomic molecules, such as oxygen and nitrogen, and one example in a polyatomic molecule, formaldehyde.

The components of the magnetic dipole moment m_x, m_y, and m_z have the symmetry species of the rotations R_x, R_y, and R_z respectively and an electronic transition between states of the same multiplicity is allowed by magnetic dipole selection rules if, analogous to equation 7.12,

$$\Gamma(\psi_e') \times \Gamma(\psi_e'') = \Gamma(R_x) \text{ and/or } \Gamma(R_y) \text{ and/or } \Gamma(R_z) \qquad (7.52)$$

for transitions between non-degenerate states or, analogous to equation 7.20,

$$\Gamma(\psi_e') \times \Gamma(\psi_e'') \supset \Gamma(R_x) \text{ and/or } \Gamma(R_y) \text{ and/or } \Gamma(R_z) \tag{7.53}$$

for transitions involving at least one degenerate state.

For example, in the case of the nitrogen molecule belonging to the $\mathbf{D}_{\infty h}$ point group, the transition

$$^1\Pi_g - {}^1\Sigma_g^+ \tag{7.54}$$

is an allowed magnetic dipole transition since $\Gamma(\psi_e') \times \Gamma(\psi_e'') = \Pi_g = \Gamma(R_{x,y})$. Such a transition has been observed. In formaldehyde, belonging to the $\mathbf{C}_{2v}$ point group, the transition

$$^1A_2 - {}^1A_1 \tag{7.55}$$

is an allowed magnetic dipole transition since $\Gamma(\psi_e') \times \Gamma(\psi_e'') = A_2 = \Gamma(R_z)$. The $\pi^* - n$ transition in formaldehyde mentioned in section 6.6 has been shown experimentally to be a weak magnetic dipole allowed $^1A_2 - {}^1A_1$ transition.

7.5 Vibrational Raman selection rules

Raman spectroscopy differs in principle from the more convential types of absorption and emission spectroscopy for which the selection rules discussed in sections 7.2-4 are appropriate.

If intense monochromatic radiation is incident on a molecular sample the light scattered by the sample will be of two types: the *Rayleigh scattering* is radiation of unchanged wavenumber while the *Raman scattering* is of radiation whose wavenumber is either greater or less than that of the incident radiation. The reason for the change in wavenumber in Raman scattering is that an electronic, vibrational or rotational transition may accompany the scattering process. Here we shall consider only vibrational transitions.

Figure 7.16 illustrates a vibrational Raman process in a simple system involving only two vibrational levels with the vibrational quantum number $v = 0$ or 1. The ratio of the populations of these two levels is given by the Boltzmann factor $\exp(-hc\tilde{\nu}/kT)$ where $\tilde{\nu}$ is the separation in cm^{-1} of the $v = 0$ and $v = 1$ levels. The intense incident radiation excites the molecules from the $v = 0$ state to a *virtual state*, a state which is not an eigenstate of the molecule, in this case V_0. From this state the molecule may emit radiation by falling down to the $v = 0$ level (Rayleigh) or to the $v = 1$ level (Raman, Stokes). Molecules which are initially in the $v = 1$ state will be excited by the incident radiation to the virtual state V_1 and may then emit radiation by falling down to the $v = 1$ level (Rayleigh) or to the $v = 0$ level

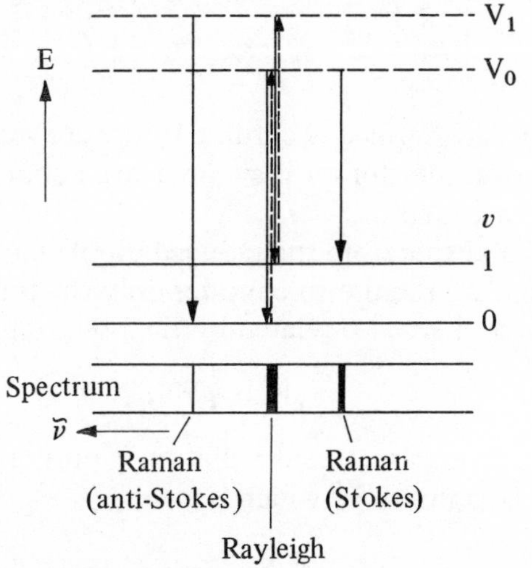

FIGURE 7.16
Transitions in a vibrational Raman spectrum

(Raman, anti-Stokes). The anti-Stokes Raman scattering is weaker than the Stokes because of lower population of the $v = 1$ level. However, we have assumed that the net transition from $v = 0$ to $v = 1$ or *vice versa* is an allowed one. Whether, in fact, it is allowed is determined by the vibrational Raman selection rules for the particular molecule and the particular vibration concerned. The general selection rule is that *a vibrational transition may occur if the transition polarizability is non-zero.*

The incident monochromatic radiation in a Raman experiment induces a dipole moment **p** in the molecule due to the electric field **E** of the radiation: **E** and **p** are vector quantities. The magnitude of this dipole moment is given by the vector equation

$$\mathbf{p} = \alpha \mathbf{E} \tag{7.56}$$

where α is the polarizability. In a molecule whose polarizability is anisotropic the direction of the induced dipole moment is different from that of the applied electric field: therefore the polarizability is a tensor and equation 7.56 can be written

$$\begin{bmatrix} p_x \\ p_y \\ p_z \end{bmatrix} = \begin{bmatrix} \alpha_{xx} & \alpha_{xy} & \alpha_{xz} \\ \alpha_{yx} & \alpha_{yy} & \alpha_{yz} \\ \alpha_{zx} & \alpha_{zy} & \alpha_{zz} \end{bmatrix} \begin{bmatrix} E_x \\ E_y \\ E_z \end{bmatrix} \tag{7.57}$$

The x-, y-, and z-axes are defined according to the conventions adopted in chapter 4. The polarizability tensor is symmetrical about the leading diagonal i.e. $\alpha_{zx} = \alpha_{xz}$, $\alpha_{yx} = \alpha_{xy}$, and $\alpha_{zy} = \alpha_{yz}$.

The electric field and therefore the induced dipole moment of equation 7.56 is oscillating and we require to consider only the time-independent factors $\mathbf{E}^0$ and $\mathbf{p}^0$ which are also related by the polarizability

$$\mathbf{p}^0 = \alpha \mathbf{E}^0 \tag{7.58}$$

The intensity of the Raman-scattered radiation is proportional to the square of the transition polarizability $\mathbf{P}^0$ which is given by

$$\mathbf{P}^0 = \int \psi_v'^* \, \mathbf{p}^0 \psi_v'' \mathrm{d}\tau \tag{7.59}$$

But from equation 7.57 it follows that

$$p_x = \alpha_{xx} E_x + \alpha_{xy} E_y + \alpha_{xz} E_z \tag{7.60}$$

$$p_y = \alpha_{xy} E_x + \alpha_{yy} E_y + \alpha_{yz} E_z \tag{7.61}$$

$$p_z = \alpha_{xz} E_x + \alpha_{yz} E_y + \alpha_{zz} E_z \tag{7.62}$$

and therefore, from equations 7.58 and 7.59,

$$p_x^0 = E_x^0 \int \psi_v'^* \alpha_{xx} \psi_v'' \mathrm{d}\tau + E_y^0 \int \psi_v'^* \alpha_{xy} \psi_v'' \mathrm{d}\tau + E_z^0 \int \psi_v'^* \alpha_{xz} \psi_v'' \mathrm{d}\tau \tag{7.63}$$

$$p_y^0 = E_x^0 \int \psi_v'^* \alpha_{xy} \psi_v'' \mathrm{d}\tau + E_y^0 \int \psi_v'^* \alpha_{yy} \psi_v'' \mathrm{d}\tau + E_z^0 \int \psi_v'^* \alpha_{yz} \psi_v'' \mathrm{d}\tau \tag{7.64}$$

$$p_z^0 = E_x^0 \int \psi_v'^* \alpha_{xz} \psi_v'' \mathrm{d}\tau + E_y^0 \int \psi_v'^* \alpha_{yz} \psi_v'' \mathrm{d}\tau + E_z^0 \int \psi_v'^* \alpha_{zz} \psi_v'' \mathrm{d}\tau \tag{7.65}$$

For the vibrational transition to be allowed the transition polarizability $\mathbf{P}^0$ of equation 7.59 must be non-zero and this requirement is fulfilled if *any* of the nine integrals of equations 7.63-65 is non-zero. For one of the integrals to be non-zero the product of the symmetry species of the quantities in the integral must, for transitions between non-degenerate vibrational levels, give the totally symmetric species, that is

$$\Gamma(\psi_v') \times \Gamma(\alpha_{ij}) \times \Gamma(\psi_v'') = A \qquad (7.66)$$

where i and j can be x, y, or z, or, for transitions between vibrational levels at least one of which is degenerate,

$$\Gamma(\psi_v') \times \Gamma(\alpha_{ij}) \times \Gamma(\psi_v'') \supset A \qquad (7.67)$$

If the lower vibrational level is the zero-point level, as is usually the case, we arrive at the vibrational Raman selection rule which states that, for a vibrational transition to be allowed,

$$\Gamma(\psi_v') = \Gamma(\alpha_{ij}) \qquad (7.68)$$

where α_{ij} represents any of the six components of the polarizability. As we shall see later, in degenerate point groups symmetry species may not be assigned to some of the components as they are written in equation 7.57, but only to linear combinations of them.

The symmetry species of the components of the polarizability are given in all the point group character tables in tables 4.1-45. We shall now see how these species are obtained in both non-degenerate and degenerate point groups.

7.5.1 NON-DEGENERATE POINT GROUPS

If the electric field is applied along the z-axis then $E_x = E_y = 0$ but $E_z \neq 0$. Therefore, from equations 7.60-62,

$$p_x = \alpha_{xz} E_z \qquad (7.69)$$

$$p_y = \alpha_{yz} E_z \qquad (7.70)$$

$$p_z = \alpha_{zz} E_z \qquad (7.71)$$

Since $\Gamma(p_x) = \Gamma(T_x)$, $\Gamma(p_y) = \Gamma(T_y)$, $\Gamma(p_z) = \Gamma(T_z)$, and $\Gamma(E_z) = \Gamma(T_z)$ it follows that

$$\Gamma(\alpha_{xz}) = \Gamma(T_x) \times \Gamma(T_z) \tag{7.72}$$

$$\Gamma(\alpha_{yz}) = \Gamma(T_y) \times \Gamma(T_z) \tag{7.73}$$

$$\Gamma(\alpha_{zz}) = \Gamma(T_z) \times \Gamma(T_z) \tag{7.74}$$

Similarly, by considering the field applied along the x- and y-axes successively it can be shown that

$$\Gamma(\alpha_{xx}) = \Gamma(T_x) \times \Gamma(T_x) \tag{7.75}$$

$$\Gamma(\alpha_{yy}) = \Gamma(T_y) \times \Gamma(T_y) \tag{7.76}$$

$$\Gamma(\alpha_{xy}) = \Gamma(T_x) \times \Gamma(T_y) \tag{7.77}$$

and therefore that, in general

$$\Gamma(\alpha_{ij}) = \Gamma(T_i) \times \Gamma(T_j) \tag{7.78}$$

Equation 7.78 illustrates the fact that Raman scattering can be regarded as involving two electric dipole allowed transitions, the first from the initial state to the virtual state and the second from the virtual state to the final state so that the selection rules are the same as for two successive electric dipole transitions.

It is clear from equations 7.74-76 that $\Gamma(\alpha_{xx})$, $\Gamma(\alpha_{yy})$, and $\Gamma(\alpha_{zz})$ are always the totally symmetric species.

In the $\mathbf{C_{2v}}$ point group, as an example of the application of equation 7.78 we get

$$\Gamma(\alpha_{xz}) = B_1 \times A_1 = B_1 \tag{7.79}$$

$$\Gamma(\alpha_{yz}) = B_2 \times A_1 = B_2 \tag{7.80}$$

$$\Gamma(\alpha_{zz}) = A_1 \times A_1 = A_1 \tag{7.81}$$

$$\Gamma(\alpha_{xx}) = B_1 \times B_1 = A_1 \tag{7.82}$$

$$\Gamma(\alpha_{yy}) = B_2 \times B_2 = A_1 \tag{7.83}$$

$$\Gamma(\alpha_{xy}) = B_1 \times B_2 = A_2 \tag{7.84}$$

The symmetry species of the α_{ij} in the $\mathbf{C}_{2v}$ point group are indicated in the final column of the character table in table 4.11.

In the $\mathbf{C}_{2h}$ point group, equation 7.78 gives

$$\Gamma(\alpha_{xz}) = B_u \times A_u = B_g \tag{7.85}$$

$$\Gamma(\alpha_{yz}) = B_u \times A_u = B_g \tag{7.86}$$

$$\Gamma(\alpha_{zz}) = A_u \times A_u = A_g \tag{7.87}$$

$$\Gamma(\alpha_{xx}) = B_u \times B_u = A_g \tag{7.88}$$

$$\Gamma(\alpha_{yy}) = B_u \times B_u = A_g \tag{7.89}$$

$$\Gamma(\alpha_{xy}) = B_u \times B_u = A_g \tag{7.90}$$

These symmetry species of α_{ij}, are indicated in the $\mathbf{C}_{2h}$ character table in table 4.22. It is noteworthy that in this point group all vibrations which are Raman-active in single quanta are 'g' vibrations, while all those which are infrared-active are 'u' vibrations. This rule is general for all point groups in which there is a centre of symmetry and is obvious in non-degenerate point groups since all translational symmetry species are 'u' species and u x u = g.

Selection rules for transitions involving a lower state which is vibrationally excited can be derived from equation 7.66 from which it follows that

$$\Gamma(\psi'_v) \times \Gamma(\psi''_v) = \Gamma(\alpha_{ij}) \tag{7.91}$$

For example in ethylene, belonging to the $\mathbf{D}_{2h}$ point group, a transition between a B_{1u} and a B_{2u} vibrational state is allowed by Raman selection rules since $B_{2u} \times B_{1u} = B_{3g} = \Gamma(\alpha_{yz})$ as can be verified using the character table in table 4.32.

7.5.2 DEGENERATE POINT GROUPS

In a degenerate point group with a unique axis (z-axis), that is all those in tables 4.1-45 except $\mathbf{T}$, $\mathbf{T}_d$, $\mathbf{O}$, $\mathbf{O}_h$, and $\mathbf{K}_h$, if the electric field is applied along the z-axis then $E_x = E_y = 0$ but $E_z \neq 0$: therefore equations 7.69-71 apply, from which it follows again that equations 7.72-74 apply and $\Gamma(\alpha_{xz})$, $\Gamma(\alpha_{yz})$, and $\Gamma(\alpha_{zz})$ are easy to obtain. However, the x- and y-axes are indistinguishable and if $E_z = 0$ and $E_x \neq 0$ then $E_y \neq 0$ also. Then, from equations 7.60-61 we get

$$p_x = \alpha_{xx} E_x + \alpha_{xy} E_y \tag{7.92}$$

$$p_y = \alpha_{xy} E_x + \alpha_{yy} E_y \tag{7.93}$$

Now consider the effect of a rotation by $2\pi/n$ about the z-axis which is a C_n axis. The effect on p_x and p_y is that

$$p_x \xrightarrow{C_n} p'_x = \alpha'_{xx} E'_x + \alpha'_{xy} E'_y \tag{7.94}$$

$$p_y \xrightarrow{C_n} p'_y = \alpha'_{xy} E'_x + \alpha'_{yy} E'_y \tag{7.95}$$

But, by analogy with the behaviour of a degenerate normal co-ordinate under the operation C_n given in equation 4.12

$$E_x \xrightarrow{C_n} E'_x = E_x \cos(2\pi/n) + E_y \sin(2\pi/n) \tag{7.96}$$

$$E_y \xrightarrow{C_n} E'_y = -E_x \sin(2\pi/n) + E_y \cos(2\pi/n) \tag{7.97}$$

(Compared with equation 4.12, $l = 1$ in the case of E_x and E_y which always have translational symmetry species for which $l = 1$). Similarly

$$p_x \xrightarrow{C_n} p'_x = p_x \cos(2\pi/n) + p_y \sin(2\pi/n) \tag{7.98}$$

$$p_y \xrightarrow{C_n} p'_y = -p_x \sin(2\pi/n) + p_y \cos(2\pi/n) \tag{7.99}$$

From equations 7.98, 7.94, 7.92, 7.93, 7.96, and 7.97

$$(\alpha_{xx} E_x + \alpha_{xy} E_y) \cos(2\pi/n) + (\alpha_{xy} E_x + \alpha_{yy} E_y) \sin(2\pi/n)$$
$$= \alpha'_{xx}(E_x \cos(2\pi/n) + E_y \sin(2\pi/n)) + \alpha'_{xy}(-E_x \sin(2\pi/n) + E_y \cos(2\pi/n)) \tag{7.100}$$

Equating coefficients of E_x and then E_y gives

$$\alpha_{xx} \cos(2\pi/n) + \alpha_{xy} \sin(2\pi/n) = \alpha'_{xx} \cos(2\pi/n) - \alpha'_{xy} \sin(2\pi/n) \tag{7.101}$$

$$\alpha_{xy} \cos(2\pi/n) + \alpha_{yy} \sin(2\pi/n) = \alpha'_{xx} \sin(2\pi/n) + \alpha'_{xy} \cos(2\pi/n) \tag{7.102}$$

Also from equations 7.99, 7.95, 7.92, 7.93, 7.96, and 7.97

$$-(\alpha_{xx}E_x + \alpha_{xy}E_y)\sin(2\pi/n) + (\alpha_{xy}E_x + \alpha_{yy}E_y)\cos(2\pi/n)$$
$$= \alpha'_{xy}(E_x\cos(2\pi/n) + E_y\sin(2\pi/n)) + \alpha'_{yy}(-E_x\sin(2\pi/n) + E_y\cos(2\pi/n)) \qquad (7.103)$$

Equating coefficients of E_x and then E_y gives

$$-\alpha_{xx}\sin(2\pi/n) + \alpha_{xy}\cos(2\pi/n) = \alpha'_{xy}\cos(2\pi/n) - \alpha'_{yy}\sin(2\pi/n) \qquad (7.104)$$

$$-\alpha_{xy}\sin(2\pi/n) + \alpha_{yy}\cos(2\pi/n) = \alpha'_{xy}\sin(2\pi/n) + \alpha'_{yy}\cos(2\pi/n) \qquad (7.105)$$

From equations 7.101 and 7.102 we get

$$\alpha'_{xx} = \alpha_{xx}\cos^2(2\pi/n) + 2\alpha_{xy}\sin(2\pi/n)\cos(2\pi/n) + \alpha_{yy}\sin^2(2\pi/n) \qquad (7.106)$$

and from equations 7.104 and 7.105

$$\alpha'_{yy} = \alpha_{xx}\sin^2(2\pi/n) - 2\alpha_{xy}\sin(2\pi/n)\cos(2\pi/n) + \alpha_{yy}\cos^2(2\pi/n) \qquad (7.107)$$

From equations 7.101 and 7.102 we can also get

$$2\alpha'_{xy} = -(\alpha_{xx} - \alpha_{yy})\sin(4\pi/n) + 2\alpha_{xy}\cos(4\pi/n) \qquad (7.108)$$

From equations 7.106 and 7.107 it follows that

$$\alpha'_{xx} + \alpha'_{yy} = \alpha_{xx} + \alpha_{yy} \qquad (7.109)$$

and

$$\alpha'_{xx} - \alpha'_{yy} = (\alpha_{xx} - \alpha_{yy})\cos(4\pi/n) + 2\alpha_{xy}\sin(4\pi/n) \qquad (7.110)$$

Equation 7.109 shows that although neither α_{xx} nor α_{yy} undergoes a simple transformation under the operation C_n, the combination $\alpha_{xx} + \alpha_{yy}$ is symmetric to this operation. It can be shown that $\alpha_{xx} + \alpha_{yy}$ is symmetric to all other operations also. By comparing equations 7.110 and 7.108 with 7.96 and 7.97 it can be seen that $2\alpha_{xy}$ and $\alpha_{xx} - \alpha_{yy}$ form a degenerate pair but since, compared with equation 4.12, $l = 2$ in equations 7.110 and 7.108, $2\alpha_{xy}$ and $\alpha_{xx} - \alpha_{yy}$ transform like an E_2 species. In some degenerate point groups, however, there is no E_2 species. In point groups with a C_3

axis the E_2 species is identical to E and in point groups with a C_4 axis, $4\pi/n = \pi$ and equations 7.108 and 7.110 become

$$\alpha'_{xy} = -\alpha_{xy} \tag{7.111}$$

$$\alpha'_{xx} - \alpha'_{yy} = -(\alpha_{xx} - \alpha_{yy}) \tag{7.112}$$

So α_{xy} and $\alpha_{xx} - \alpha_{yy}$ have the species B_2 and B_1 respectively in point groups which have these species as well as a C_4 axis. Thus we have that $\Gamma(\alpha_{xy}) = B_2$ and $\Gamma(\alpha_{xx} - \alpha_{yy}) = B_1$ in the $\mathbf{D}_4$, $\mathbf{C}_{4v}$, and $\mathbf{D}_{2d}$ point groups while $\Gamma(\alpha_{xy}) = \Gamma(\alpha_{xx} - \alpha_{yy}) = B$ in the $\mathbf{C}_4$ and $\mathbf{S}_4$ point groups. Where there is a centre of symmetry the species are 'g' species, therefore, in the $\mathbf{C}_{4h}$ point group, $\Gamma(\alpha_{xy}) = \Gamma(\alpha_{xx} - \alpha_{yy}) = B_g$ and in the $\mathbf{D}_{4h}$ point group $\Gamma(\alpha_{xy}) = B_{2g}$ and $\Gamma(\alpha_{xx} - \alpha_{yy}) = B_{1g}$.

In the degenerate point groups $\mathbf{T}$, $\mathbf{T}_d$, $\mathbf{O}$, $\mathbf{O}_h$ and $\mathbf{K}_h$ which have no unique axes it is the transformation properties of α_{xy}, α_{xz}, α_{yz}, $\alpha_{xx} + \alpha_{yy} + \alpha_{zz}$, $2\alpha_{zz} - \alpha_{xx} - \alpha_{yy}$, and $\alpha_{xx} - \alpha_{yy}$ which can be assigned to symmetry species.

As an example of the application of vibrational Raman selection rules in a degenerate point group it can be seen from the $\mathbf{O}_h$ character table in table 4.43 that vibrations active in single quanta in SF_6 which belongs to the $\mathbf{O}_h$ point group are those of species a_{1g}, e_g, and t_{2g} whereas infra-red active vibrations are of species t_{1u} only.

If the vibrational transition involves a vibrationally excited lower state then it follows from equation 7.67 that

$$\Gamma(\psi'_v) \times \Gamma(\psi''_v) \supset \Gamma(\alpha_{ij}) \tag{7.113}$$

For example if, in the cyclopentadienyl radical which belongs to the $\mathbf{D}_{5h}$ point group, a vibrational transition is considered between a lower state of species e'_1 and an upper state of species e'_2 then, from table 4.52,

$$\Gamma(\psi'_v) \times \Gamma(\psi''_v) = e'_2 \times e'_1 = E'_1 + E'_2 \tag{7.114}$$

and, since $\Gamma(\alpha_{xx} - \alpha_{yy})$ and $\Gamma(\alpha_{xy}) = E'_2$, this transition is allowed by Raman selection rules.

7.6 Electric quadrupole selection rules

There are a few cases in diatomic molecules in which electronic transitions occur due to a non-zero electric quadrupole transition moment. Such transitions are extremely weak, even weaker than transitions due to a non-zero magnetic dipole transition moment.

The electric quadrupole moment is a tensor whose components have the same symmetry species as those of the polarizability tensor. Therefore the electric quadrupole selection rules are exactly the same as the Raman selection rules so that, for example, the transitions

$$\begin{aligned} &\Pi_g - \Sigma_g^+ \\ &\Delta_g - \Sigma_g^+ \end{aligned} \qquad (7.115)$$

are allowed by these selection rules in the $\mathbf{D}_{\infty h}$ point group.

7.7 Number of normal vibrations of each symmetry species

In section 1.3 it was shown that there are $3N - 5$ normal vibrations in a linear molecule and $3N - 6$ in a non-linear molecule, where N is the number of atoms in the molecule. There is a set of fairly simple rules for determining the number of vibrations belonging to each of the symmetry species of the point group to which the molecule belongs. These rules involve the concept of *sets of equivalent nuclei.* Nuclei form a set if they can be transformed into one another by any of the symmetry operations of the point group. For example in the $\mathbf{C}_{2v}$ point group there can be, as illustrated in figure 7.17, four kinds of set:

(*i*) if a nucleus, like one of those marked '1', does not lie on any element of symmetry, C_2, $\sigma_v(xz)$, or $\sigma_v(yz)$, then it must be one of a set of *four* equivalent nuclei in order that the molecule is symmetric with respect to all symmetry elements;

(*ii*) if a nucleus, like one of those marked '2', lies on the $\sigma_v(xz)$ element only then it must be one of a set of *two*;

(*iii*) if a nucleus, like one of those marked '3', lies on the $\sigma_v(yz)$ element only then it also must be one of a set of *two*;

(*iv*) if a nucleus, like that marked '4', lies on all the symmetry elements then this atom forms a set of *one.*

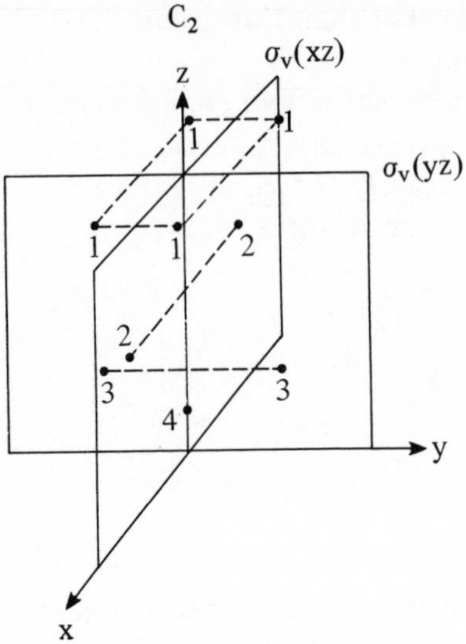

FIGURE 7.17
Illustration of the four types of sets of identical nuclei which are possible in the $\mathbf{C_{2v}}$ point group

7.1.1 NON-DEGENERATE VIBRATIONS

In the case of a non-degenerate vibration of a particular symmetry species all nuclei in a set move in exactly the same way. Each nucleus has three degrees of freedom so the nuclei of a set can contribute a maximum of three degrees of freedom to each non-degenerate species. This maximum is attained if the nuclei of the set do not lie on any symmetry element. If there are m sets of this kind they contribute $3m$ degrees of freedom to each symmetry type. But if the nuclei of a set lie on one or more symmetry elements then they contribute $2 \times m$, $1 \times m$, or $0 \times m$ degrees of freedom to a symmetry type depending on the element of symmetry and the symmetry type. Thus the total number of degrees of freedom of each symmetry species can be determined.

For example in the $\mathbf{C_{2v}}$ point group each of the four types of sets of nuclei illustrated in figure 7.17 will contribute degrees of freedom to the symmetry species as follows:

(*i*) *Set of nuclei '1'*
Each set of this type will contribute three degrees of freedom to each symmetry species. If there are m sets each will contribute $3m$ degrees of freedom to each symmetry species as indicated in table 7.4.

TABLE 7.4
Number of normal vibrations of each symmetry species in the $\mathbf{C}_{2v}$ point group

	Degrees of freedom contributed by sets of nuclei						
Species	*On no symmetry element*	*On* $\sigma_v(xz)$	*On* $\sigma_v(yz)$	*On all symmetry elements*	*Translational degrees of freedom*	*Rotational degrees of freedom*	*Number of normal vibrations*
a_1	$3m$	$2m_{xz}$	$2m_{yz}$	$1m_0$	1	0	$3m + 2m_{xz} + 2m_{yz} + m_0 - 1$
a_2	$3m$	$1m_{xz}$	$1m_{yz}$	0	0	1	$3m + m_{xz} + m_{yz} - 1$
b_1	$3m$	$2m_{xz}$	$1m_{yz}$	$1m_0$	1	1	$3m + 2m_{xz} + m_{yz} + m_0 - 2$
b_2	$3m$	$1m_{xz}$	$2m_{yz}$	$1m_0$	1	1	$3m + m_{xz} + 2m_{yz} + m_0 - 2$

(*ii*) *Set of nuclei '2'*

If the motions of nuclei of this type of set are to be of species a_1 then, in order to be symmetric to all operations they can move only in the xz-plane and have two degrees of freedom. If the motions are to be of species a_2 they must be antisymmetric to reflection in both planes and therefore the two atoms must move in opposite directions but along lines perpendicular to the xz-plane. Consequently they contribute only one degree of freedom to the a_2 species. Similarly they contribute two degrees of freedom to the b_1 and one to the b_2 species. The number of sets of nuclei of this type is designated m_{xz} and the total number of degrees of freedom contributed to the various symmetry species are given in table 7.4.

(*iii*) *Set of nuclei '3'*

The motions of nuclei in sets of this type are analogous to those discussed in (*ii*) and their assignment to symmetry species is analogous. The number of sets of this type is designated m_{yz} and assignment of the motions to symmetry species is given in table 7.4.

(*iv*) *Set of nuclei '4'*

In each set of this type there is necessarily only one nucleus. When its motion is symmetric to all operations, it can move only along the C_2 axis giving one degree of freedom belonging to the a_1 species. It cannot move in such a way as to be antisymmetric to both planes so no degrees of freedom belong to the a_2 species. For a degree of freedom to be antisymmetric to reflection in one of the planes it must move in a line perpendicular to that plane, so one degree of freedom belongs to each of the b_1 and b_2 species. The number of sets of this type is designated m_o and the assignments of the motions are given in table 7.4.

However, we know that a non-linear molecule has three rotational and three translational degrees of freedom all of which can be assigned to symmetry species (section 4.4). These are indicated in table 7.4 and subtracted from the total number of degrees of freedom to give the total number of vibrational degrees of freedom.

In table 7.5 are given formulae, analogous to those derived for the C_{2v} point group, for determining the number of normal vibrations belonging to the various symmetry species in all non-degenerate point groups.

TABLE 7.5
Number of vibrations of each species for non-degenerate point groups

Point group	*Species*	*Number of vibrations*[1]
$\mathbf{C_2}$	a	$3m + m_0 - 2$
	b	$3m + 2m_0 - 4$
$\mathbf{C_s}$	a′	$3m + 2m_0 - 3$
	a″	$3m + m_0 - 3$
$\mathbf{C_i}$	a_g	$3m - 3$
	a_u	$3m + 3m_0 - 3$
$\mathbf{C_{2v}}$	a_1	$3m + 2m_{xz} + 2m_{yz} + m_0 - 1$
	a_2	$3m + m_{xz} + m_{yz} - 1$
	b_1	$3m + 2m_{xz} + m_{yz} + m_0 - 2$
	b_2	$3m + m_{xz} + 2m_{yz} + m_0 - 2$
$\mathbf{C_{2h}}$	a_g	$3m + 2m_h + m_2 - 1$
	a_u	$3m + m_h + m_2 + m_0 - 1$
	b_g	$3m + m_h + 2m_2 - 2$
	b_u	$3m + 2m_h + 2m_2 + 2m_0 - 2$
$\mathbf{D_2}$	a	$3m + m_{2x} + m_{2y} + m_{2z}$
	b_1	$3m + 2m_{2x} + 2m_{2y} + m_{2z} + m_0 - 2$
	b_2	$3m + 2m_{2x} + m_{2y} + 2m_{2z} + m_0 - 2$
	b_3	$3m + m_{2x} + 2m_{2y} + 2m_{2z} + m_0 - 2$
$\mathbf{D_{2h}}$	a_g	$3m + 2m_{xy} + 2m_{xz} + 2m_{yz} + m_{2x} + m_{2y} + m_{2z}$
	a_u	$3m + m_{xy} + m_{xz} + m_{yz}$
	b_{1g}	$3m + 2m_{xy} + m_{xz} + m_{yz} + m_{2x} + m_{2y} - 1$
	b_{1u}	$3m + m_{xy} + 2m_{xz} + 2m_{yz} + m_{2x} + m_{2y} + m_{2z} + m_0 - 1$
	b_{2g}	$3m + m_{xy} + 2m_{xz} + m_{yz} + m_{2x} + m_{2z} - 1$
	b_{2u}	$3m + 2m_{xy} + m_{xz} + 2m_{yz} + m_{2x} + m_{2y} + m_{2z} + m_0 - 1$
	b_{3g}	$3m + m_{xy} + m_{xz} + 2m_{yz} + m_{2y} + m_{2z} - 1$
	b_{3u}	$3m + 2m_{xy} + 2m_{xz} + m_{yz} + m_{2x} + m_{2y} + m_{2z} + m_0 - 1$

[1] m is the number of sets of equivalent nuclei not on any symmetry element; m_{xy}, m_{xz}, m_{yz} are the number of sets lying on the xy, xz, yz, planes respectively but not on any axes going through these planes; m_2 is the number of sets of nuclei on a C_2 axis but not at the point of intersection with any other symmetry element; m_{2x}, m_{2y}, m_{2z} are the number of sets lying on the x, y, or z axes respectively if they are two-fold axes but not on all of them; m_h is the number of sets lying on the σ_h plane but not on the axis perpendicular to it; m_0 is the number of nuclei lying on all symmetry elements.

As an example of the use of these formulae we choose naphthalene for which, using the axis notation in figure 4.4, $m_{yz} = 4$, $m_{2z} = 1$, and all other m's are zero. This gives the result that the forty-eight normal vibrations are distributed as follows: $9a_g$, $4a_u$, $3b_{1g}$, $8b_{1u}$, $4b_{2g}$, $8b_{2u}$, $8b_{3g}$, $4b_{3u}$.

7.7.2 DEGENERATE VIBRATIONS

In a molecule belonging to a degenerate point group, for example $\mathbf{C_{3v}}$, the non-degenerate vibrations of the various sets of equivalent nuclei can be treated as in section 7.7.1.

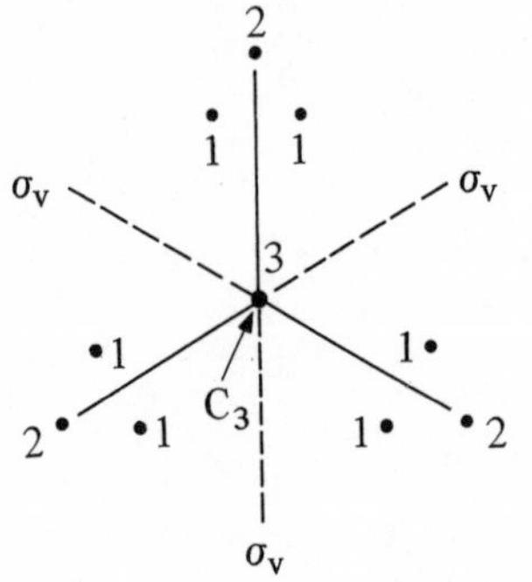

FIGURE 7.18
Illustration of the three types of sets of identical nuclei which are possible in the $\mathbf{C_{3v}}$ point group

In the $\mathbf{C_{3v}}$ point group there are three types of sets of nuclei. These are illustrated in figure 7.18 which shows a view down the C_3 axis.

(*i*) *Set of nuclei '1'*
These nuclei lie on no element of symmetry and there must be six nuclei belonging to each set. The six nuclei have eighteen degrees of freedom of which three belong to each of the a_1 and a_2 species. In a doubly degenerate degree of freedom a displacement of one of the nuclei of a set corresponds to *two* different displacements of each of the other nuclei of the set. Therefore the set of nuclei '1' has twelve degrees of freedom comprising six doubly degenerate degrees of freedom. If there are m such sets they contribute $6m$ doubly degenerate degrees of freedom.

(*ii*) *Set of nuclei '2'*
These nuclei lie on the σ_v planes of symmetry and there are three nuclei belonging to each set. The three nuclei have nine degrees of freedom of

which two belong to the a_1 species and one to the a_2 species. The nuclei have six degenerate degrees of freedom since a displacement of one nucleus corresponds to two different displacements of the other two nuclei. These six degrees of freedom comprise three doubly degenerate degrees of freedom. If there are m_v such sets they contribute $3m_v$ doubly degenerate degrees of freedom.

(*iii*) *Set of nuclei '3'*

There is only one nucleus in this set since it must lie on all elements of symmetry. Of the three degrees of freedom one, along the C_3 axis, belongs to the a_1 species and the other two are perpendicular to each other and to the C_3 axis: they are obviously degenerate so two degrees of freedom comprise one doubly degenerate degree of freedom. If there are m_o such sets they contribute m_o doubly degenerate degrees of freedom.

In table 7.6 the results for the $\mathbf{C}_{3v}$ point group are summarized and the translational and rotational degrees of freedom are subtracted to give, in the final column, the number of vibrations of each symmetry species.

TABLE 7.6
Number of normal vibrations of each symmetry species in the $\mathbf{C}_{3v}$ point group

	Degrees of freedom contributed by sets of nuclei					
Species	*On no symmetry element*	*On* σ_v	*On all symmetry elements*	*Translational degrees of freedom*	*Rotational degrees of freedom*	*Number of normal vibrations*
a_1	$3m$	$2m_v$	$1m_0$	1	0	$3m + 2m_v + m_0 - 1$
a_2	$3m$	m_v	0	0	1	$3m + m_v - 1$
e	$6m$	$3m_v$	$1m_0$	1	1	$6m + 3m_v + m_0 - 2$

Degenerate vibrations in other point groups can be treated in an analogous way. Table 7.7 lists the formulae for calculating the number of degenerate and non-degenerate vibrations in molecules belonging to degenerate point groups.

As an example, for benzene $m_2 = 2$ and all the other m's are zero which

TABLE 7.7
Number of vibrations of each species for degenerate point groups

Point group	*Species*	*Number of vibrations*[1]
$\mathbf{C_3}$	a	$3m + m_0 - 2$
	e	$3m + m_0 - 2$
$\mathbf{C_4}$	a	$3m + m_0 - 2$
	b	$3m$
	e	$3m + m_0 - 2$
$\mathbf{C_6}$	a	$3m + m_0 - 2$
	b	$3m$
	e_1	$3m + m_0 - 2$
	e_2	$3m$
$\mathbf{S_4}$	a	$3m + m_2 - 1$
	b	$3m + m_2 + m_0 - 1$
	e	$3m + 2m_2 + m_0 - 2$
$\mathbf{S_6}$	a_g	$3m + m_3 - 1$
	b_u	$3m + m_3 + m_0 - 1$
	e_g	$3m + m_3 - 1$
	e_u	$3m + m_3 + m_0 - 1$
$\mathbf{D_3}$	a_1	$3m + m_2 + m_3$
	a_2	$3m + 2m_2 + m_3 + m_0 - 2$
	e	$6m + 3m_2 + 2m_3 + m_0 - 2$
$\mathbf{D_4}$	a_1	$3m + m_2 + m_2' + m_4$
	a_2	$3m + 2m_2 + 2m_2' + m_4 + m_0 - 2$
	b_1	$3m + m_2 + 2m_2'$
	b_2	$3m + 2m_2 + m_2'$
	e	$6m + 3m_2 + 3m_2' + 2m_4 + m_0 - 2$

TABLE 7.7 continued

Point group	*Species*	*Number of vibrations*[1]
$\mathbf{D_6}$	a_1	$3m + m_2 + m_2' + m_6$
	a_2	$3m + 2m_2 + 2m_2' + m_6 + m_0 - 2$
	b_1	$3m + m_2 + 2m_2'$
	b_2	$3m + 2m_2 + m_2'$
	e_1	$6m + 3m_2 + 3m_2' + 2m_6 + m_0 - 2$
	e_2	$6m + 3m_2 + 3m_2'$
$\mathbf{C_{3v}}$	a_1	$3m + 2m_v + m_0 - 1$
	a_2	$3m + m_v - 1$
	e	$6m + 3m_v + m_0 - 2$
$\mathbf{C_{4v}}$	a_1	$3m + 2m_v + 2m_d + m_0 - 1$
	a_2	$3m + m_v + m_d - 1$
	b_1	$3m + 2m_v + m_d$
	b_2	$3m + m_v + 2m_d$
	e	$6m + 3m_v + 3m_d + m_0 - 2$
$\mathbf{C_{5v}}$	a_1	$3m + 2m_v + m_0 - 1$
	a_2	$3m + m_v - 1$
	e_1	$6m + 3m_v + m_0 - 2$
	e_2	$6m + 3m_v$
$\mathbf{C_{6v}}$	a_1	$3m + 2m_v + 2m_d + m_0 - 1$
	a_2	$3m + m_v + m_d - 1$
	b_1	$3m + 2m_v + m_d$
	b_2	$3m + m_v + 2m_d$
	e_1	$6m + 3m_v + 3m_d + m_0 - 2$
	e_2	$6m + 3m_v + 3m_d$

TABLE 7.7 continued

Point group	*Species*	*Number of vibrations*[1]
$\mathbf{C_{\infty v}}$	σ^+	$m_0 - 1$
	σ^-	0
	π	$m_0 - 2$
	$\delta, \phi \ldots$	0
$\mathbf{C_{3h}}$	a′	$3m + 2m_h + m_3 - 1$
	a″	$3m + m_h + m_3 + m_0 - 1$
	e′	$3m + 2m_h + m_3 + m_0 - 1$
	e″	$3m + m_h + m_3 - 1$
$\mathbf{C_{4h}}$	a_g	$3m + 2m_h + m_4 - 1$
	a_u	$3m + m_h + m_4 + m_0 - 1$
	b_g	$3m + 2m_h$
	b_u	$3m + m_h$
	e_g	$3m + m_h + m_4 - 1$
	e_u	$3m + 2m_h + m_4 + m_0 - 1$
$\mathbf{C_{6h}}$	a_g	$3m + 2m_h + m_6 - 1$
	a_u	$3m + m_h + m_6 + m_0 - 1$
	b_g	$3m + m_h$
	b_u	$3m + 2m_h$
	e_{1g}	$3m + m_h + m_6 - 1$
	e_{1u}	$3m + 2m_h + m_6 + m_0 - 1$
	e_{2g}	$3m + 2m_h$
	e_{2u}	$3m + m_h$
$\mathbf{D_{2d}}$	a_1	$3m + 2m_d + m_2 + m_4$
	a_2	$3m + m_d + 2m_2 - 1$
	b_1	$3m + m_d + m_2$
	b_2	$3m + 2m_d + 2m_2 + m_4 + m_0 - 1$
	e	$6m + 3m_d + 3m_2 + 2m_4 + m_0 - 2$

TABLE 7.7 continued

Point group	*Species*	*Number of vibrations*[1]
$\mathbf{D_{3d}}$	a_{1g}	$3m + 2m_d + m_2 + m_6$
	a_{1u}	$3m + m_d + m_2$
	a_{2g}	$3m + m_d + 2m_2 - 1$
	a_{2u}	$3m + 2m_d + 2m_2 + m_6 + m_0 - 1$
	e_g	$6m + 3m_d + 3m_2 + m_6 - 1$
	e_u	$6m + 3m_d + 3m_2 + m_6 + m_0 - 1$
$\mathbf{D_{4d}}$	a_1	$3m + 2m_d + m_2 + m_8$
	a_2	$3m + m_d + 2m_2 - 1$
	b_1	$3m + m_d + m_2$
	b_2	$3m + 2m_d + 2m_2 + m_8 + m_0 - 1$
	e_1	$6m + 3m_d + 3m_2 + m_8 + m_0 - 1$
	e_2	$6m + 3m_d + 3m_2$
	e_3	$6m + 3m_d + 3m_2 + m_8 - 1$
$\mathbf{D_{3h}}$	a_1'	$3m + 2m_v + 2m_h + m_2 + m_3$
	a_1''	$3m + m_v + m_h$
	a_2'	$3m + m_v + 2m_h + m_2 - 1$
	a_2''	$3m + 2m_v + m_h + m_2 + m_3 + m_0 - 1$
	e'	$6m + 3m_v + 4m_h + 2m_2 + m_3 + m_0 - 1$
	e''	$6m + 3m_v + 2m_h + m_2 + m_3 - 1$
$\mathbf{D_{4h}}$	a_{1g}	$3m + 2m_v + 2m_d + 2m_h + m_2 + m_2' + m_4$
	a_{1u}	$3m + m_v + m_d + m_h$
	a_{2g}	$3m + m_v + m_d + 2m_h + m_2 + m_2' - 1$
	a_{2u}	$3m + 2m_v + 2m_d + m_h + m_2 + m_2' + m_4 + m_0 - 1$
	b_{1g}	$3m + 2m_v + m_d + 2m_h + m_2 + m_2'$
	b_{1u}	$3m + m_v + 2m_d + m_h + m_2'$
	b_{2g}	$3m + m_v + 2m_d + 2m_h + m_2 + m_2'$
	b_{2u}	$3m + 2m_v + m_d + m_h + m_2$
	e_g	$6m + 3m_v + 3m_d + 2m_h + m_2 + m_2' + m_4 - 1$
	e_u	$6m + 3m_v + 3m_d + 4m_h + 2m_2 + 2m_2' + m_4 + m_0 - 1$

TABLE 7.7 continued

Point group	*Species*	*Number of vibrations*[1]
$\mathbf{D_{5h}}$	a_1'	$3m + 2m_v + 2m_h + m_2 + m_5$
	a_1''	$3m + m_v + m_h$
	a_2'	$3m + m_v + 2m_h + m_2 - 1$
	a_2''	$3m + 2m_v + m_h + m_2 + m_5 + m_0 - 1$
	e_1'	$6m + 3m_v + 4m_h + 2m_2 + m_5 + m_0 - 1$
	e_1''	$6m + 3m_v + 2m_h + m_2 + m_5 - 1$
	e_2'	$6m + 3m_v + 4m_h + 2m_2$
	e_2''	$6m + 3m_v + 2m_h + m_2$
$\mathbf{D_{6h}}$	a_{1g}	$3m + 2m_v + 2m_d + 2m_h + m_2 + m_2' + m_6$
	a_{1u}	$3m + m_v + m_d + m_h$
	a_{2g}	$3m + m_v + m_d + 2m_h + m_2 + m_2' - 1$
	a_{2u}	$3m + 2m_v + 2m_d + m_h + m_2 + m_2' + m_6 + m_0 - 1$
	b_{1g}	$3m + m_v + 2m_d + m_h + m_2'$
	b_{1u}	$3m + 2m_v + m_d + 2m_h + m_2 + m_2'$
	b_{2g}	$3m + 2m_v + m_d + m_h + m_2$
	b_{2u}	$3m + m_v + 2m_d + 2m_h + m_2 + m_2'$
	e_{1g}	$6m + 3m_v + 3m_d + 2m_h + m_2 + m_2' + m_6 - 1$
	e_{1u}	$6m + 3m_v + 3m_d + 4m_h + 2m_2 + 2m_2' + m_6 + m_0 - 1$
	e_{2g}	$6m + 3m_v + 3m_d + 4m_h + 2m_2 + 2m_2'$
	e_{2u}	$6m + 3m_v + 3m_d + 2m_h + m_2 + m_2'$
$\mathbf{D_{\infty h}}$	σ_g^+	m_∞
	σ_u^+	$m_\infty + m_0 - 1$
	σ_g^-, σ_u^-	0
	π_g	$m_\infty - 1$
	π_u	$m_\infty + m_0 - 1$
	$\delta_g, \delta_u, \phi_g, \phi_u \ldots$	0

TABLE 7.7 continued

Point group	*Species*	*Number of vibrations*[1]
T	a	$3m + m_2 + m_3$
	e	$3m + m_2 + m_3$
	t	$9m + 5m_2 + 3m_3 + m_0 - 2$
$\mathbf{T_d}$	a_1	$3m + 2m_d + m_2 + m_3$
	a_2	$3m + m_d$
	e	$6m + 3m_d + m_2 + m_3$
	t_1	$9m + 4m_d + 2m_2 + m_3 - 1$
	t_2	$9m + 5m_d + 3m_2 + 2m_3 + m_0 - 1$
$\mathbf{O_h}$	a_{1g}	$3m + 2m_h + 2m_d + m_2 + m_3 + m_4$
	a_{1u}	$3m + m_h + m_d$
	a_{2g}	$3m + 2m_h + m_d + m_2$
	a_{2u}	$3m + m_h + 2m_d + m_2 + m_3$
	e_g	$6m + 4m_h + 3m_d + 2m_2 + m_3 + m_4$
	e_u	$6m + 2m_h + 3m_d + m_2 + m_3$
	t_{1g}	$9m + 4m_h + 4m_d + 2m_2 + m_3 + m_4 - 1$
	t_{1u}	$9m + 5m_h + 5m_d + 3m_2 + 2m_3 + 2m_4 + m_0 - 1$
	t_{2g}	$9m + 4m_h + 5m_d + 2m_2 + 2m_3 + m_4$
	t_{2u}	$9m + 5m_h + 4m_d + 2m_2 + m_3 + m_4$

[1] m is the number of sets of equivalent nuclei not on any symmetry element; m_2, m_3, . . . are the number of sets on two-, three-, . . . fold axes but not on any other symmetry element which does not wholly coincide with the axis; m_2' is the number of sets on a C_2' axis (see character tables); m_v, m_d, m_h are the number of sets on σ_v, σ_d, σ_h planes respectively but not on any other symmetry element; m_0 is the number of nuclei lying on all symmetry elements.

gives the result that the thirty normal vibrations are distributed as follows: $2a_{1g}$, $0a_u$, $1a_{2g}$, $1a_{2u}$, $0b_{1g}$, $2b_{1u}$, $2b_{2g}$, $2b_{2u}$, $1e_{1g}$, $3e_{1u}$, $4e_{2g}$, $2e_{2u}$ (each doubly degenerate vibration counts as two vibrations).

7.8 Potential energy curves with more than one minimum

7.8.1 POTENTIAL ENERGY CURVES FOR INVERSION AND HYDROGEN-BONDING VIBRATIONS

7.8.1(a) Inversion Vibrations

It was mentioned briefly in section 1.1 that ammonia has a pyramidal configuration in its ground electronic state and therefore belongs to the $\mathbf{C}_{3v}$ point group but that it is planar in several of its excited electronic states in which it belongs to the $\mathbf{D}_{3h}$ point group. So far as the applications of molecular symmetry are concerned the problems which this situation poses are:

(*i*) In an electronic transition between a non-planar and a planar configuration how do we determine the electronic and vibronic selection rules?

(*ii*) How do we correlate both the electronic and vibronic energy levels as we proceed smoothly from the non-planar to the planar configuration?

The answers to both these problems lie in a detailed examination of the potential energy curve for the normal vibration which takes the molecule from one configuration to the other. In the case of ammonia this is the vibration ν_2 illustrated in figure 4.5. If we plot the potential energy curve in the ground electronic state for this vibration it will not have a single minimum like the one in figure 1.12 but two minima of equal energy corresponding to the two energetically equivalent configurations (i) and (iii) shown in figure 7.19. The planar configuration (ii) is unstable and produces a barrier in the potential energy curve which, regarding the situation classically rather than quantum mechanically, would have to be surmounted in order to go from configuration (i) to (iii), a process called *inversion*: ν_2 is called the *inversion vibration.* The height of the barrier is 2076 cm^{-1} in ammonia and the potential energy curve is rather like that in figure 7.20 for an intermediate barrier. Figure 7.20 shows also the potential energy curve for the extreme cases of zero barrier (a planar molecule) and an infinite barrier in which the molecule cannot undergo inversion. The vibrational energy levels are harmonic in the two extreme cases but, as the infinite barrier is reduced on going from right to left of figure 7.20, what happens quantum mechanically is that tunnelling through the barrier occurs. The vibrational

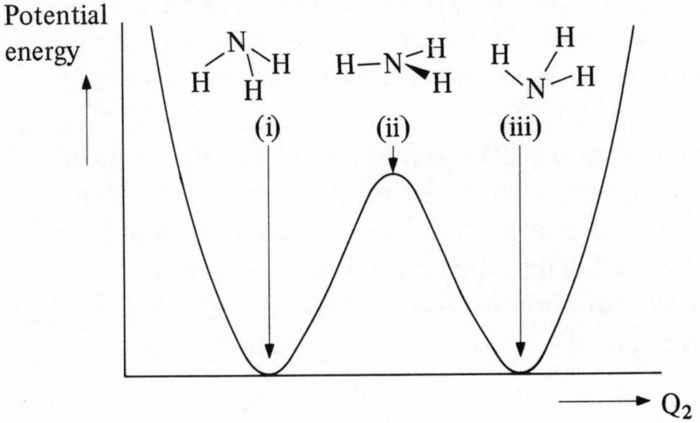

FIGURE 7.19
Double minimum potential function for the inversion vibration (ν_2) of ammonia

energy levels are split into pairs, the splitting being larger nearer the top of the barrier. But, as shown in the curve for an intermediate barrier, the effect of the barrier on the energy levels is to make them anharmonic

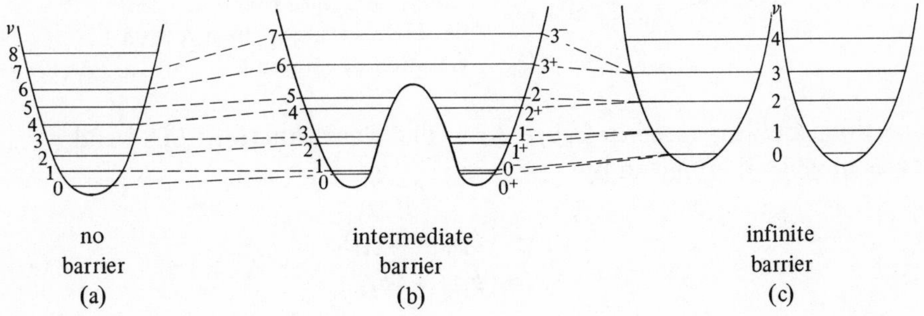

FIGURE 7.20
Correlation of inversion vibration energy levels from the case (a) where there is no barrier to the case (c) where the barrier is infinite

even above the barrier. The splitting of the levels depends on the height and width of the barrier and also on the masses of the atoms involved in the inversion. For example, the splitting of the zero-point level is 0·793 cm^{-1} in NH_3 and 0·053 cm^{-1} in ND_3 which illustrates how rapidly tunnelling decreases with increasing mass.

The method of numbering the vibrational levels in the case of the intermediate barrier presents a choice according to the context of its use. If it is appropriate to make a comparison with the zero barrier case then the v = 0, 1, 2, 3 . . . numbering given on the left hand side of the intermediate barrier curve in figure 7.20 should be used but if it is more appropriate to make the comparison with the infinite barrier case then the v = 0^+, 0^-, 1^+, 1^-, . . . numbering system given on the right hand side of the curve should be used. The superscript + or – refers to whether the vibrational wave function is respectively symmetric or antisymmetric to inversion. The 0^+, 0^-, 1^+, and 1^- vibrational wave

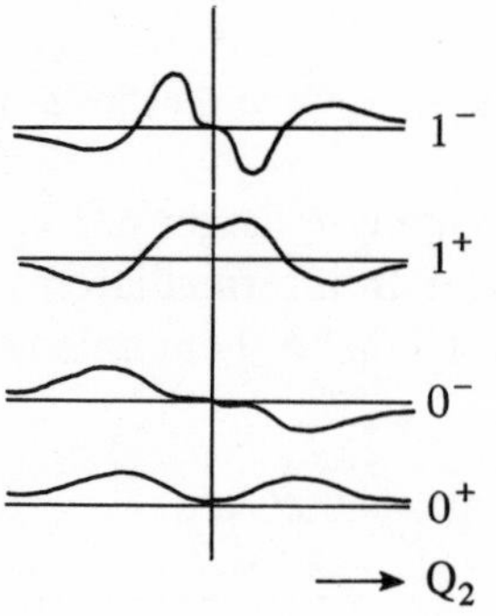

FIGURE 7.21
Vibrational wave functions for 0^+, 0^-, 1^+, 1^- levels of the inversion vibration of ammonia

functions are illustrated in figure 7.21. The wave functions of ν^+ and ν^- levels in general are given by

$$\begin{aligned} \psi_{\nu^+} &= (\psi_{\mathrm{I}})_\nu + (\psi_{\mathrm{III}})_\nu \\ \psi_{\nu^-} &= (\psi_{\mathrm{I}})_\nu - (\psi_{\mathrm{III}})_\nu \end{aligned} \qquad (7.116)$$

where $(\psi_{\mathrm{I}})_\nu$ and $(\psi_{\mathrm{III}})_\nu$ are the vibrational wave functions for the equivalent forms (i) and (iii) of figure 7.19 for the case of an infinite barrier.

The additional property of symmetry or antisymmetry to inversion when the barrier is low enough for energy levels to be split means that effectively a σ_h plane of symmetry has been introduced in addition to the symmetry elements of the $\mathbf{C_{3v}}$ point group. Therefore the vibrational levels should be classified according to the symmetry species of the $\mathbf{D_{3h}}$ point group.

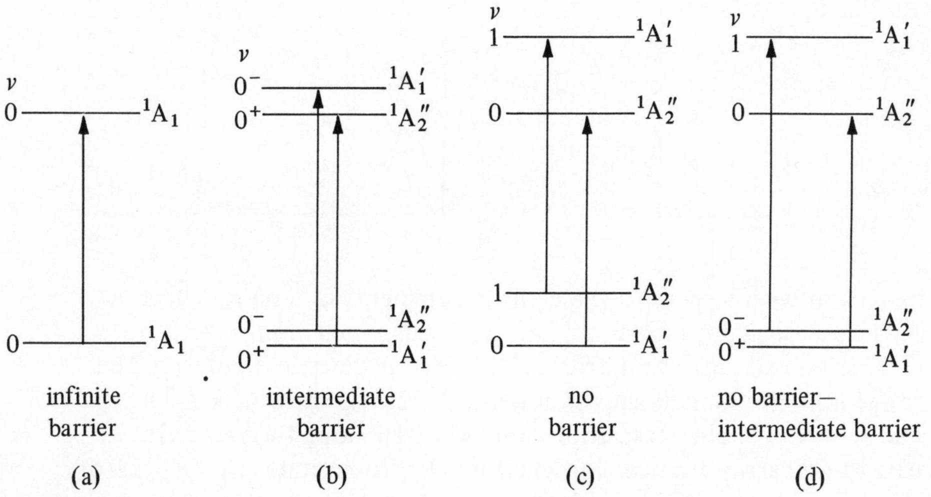

FIGURE 7.22
Illustration of four cases of the splitting of a $^1A_1 - {}^1A_1$ electronic transition in ammonia. The order in the excited state of the $^1A_1'$, $^1A_2''$ pairs has been chosen to be opposite to that in the ground state.

Figure 7.22(a) to (c) shows what happens, as the barrier is lowered, to an electronic transition which is $^1A_1 - {}^1A_1$ in a molecule belonging to the $\mathbf{C_{3v}}$ point group starting with an infinite barrier to inversion in both electronic states. The $^1A_1 - {}^1A_1$ transition in figure 7.22(a) is allowed by electric dipole selection rules and is polarized along the z-axis. When the levels are split by different amounts in both electronic states (figure 7.22(b)) the $^1A_1 - {}^1A_1$ transition splits into two allowed transitions. Table 7.8 shows how the symmetry species of the $\mathbf{C_{3v}}$ point group correlate with those of the $\mathbf{D_{3h}}$ point group by preserving their

TABLE 7.8
Correlation of symmetry species in the $\mathbf{C_{3v}}$ and $\mathbf{D_{3h}}$ point groups

$\mathbf{C_{3v}}$	$2C_3$	$3\sigma_v$
A_1	1	1
A_2	1	−1
E	−1	0

$\mathbf{D_{3h}}$	$2C_3$	$3\sigma_v$	σ_h
A_1'	1	1	1
A_2'	1	−1	1
E'	−1	0	2
A_1''	1	−1	−1
A_2''	1	1	−1
E''	−1	0	−2

behaviour with respect to the common elements C_3 and σ_v. Thus the A_1 levels each split into an A_1' and A_2'' level, which are respectively symmetric and antisymmetric to σ_h. The only electric dipole allowed transitions are the two shown in figure 7.22(b)). Since $A_1' \times A_2'' = A_2''$ and $A_2'' = \Gamma(T_z)$ both transitions are polarized along the z-axis. In the case of no barrier to inversion in either electronic state (figure 7.22(c)) the two transitions become, in the case illustrated, 0 – 0 and 1 – 1 transitions in ν_2 in the planar molecule. Figure 7.22(d)) illustrates the case of a non-planar ground state with an intermediate barrier to inversion and a planar excited state.

Figure 7.23 illustrates a vibrational transition in the ground electronic state of ammonia. The transition is a 1 – 0 transition in ν_3 which is, in the

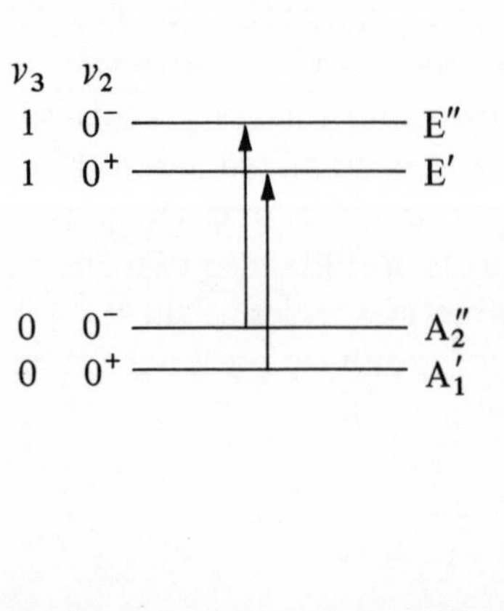

FIGURE 7.23
The splitting due to inversion in ammonia of a $\nu = 0$ to $\nu = 1$ vibrational transition in the doubly degenerate vibration ν_3.

$\mathbf{C_{3v}}$ point group, an e species vibration (figure 4.5). Each of the levels is doubled by inversion and the components are classified in figure 7.23 according to the $\mathbf{D_{3h}}$ point group. The two allowed transitions are $E' - A_1'$ and $E'' - A_2''$ both polarized in the xy-plane. Since the inversion doubling is similar for $\nu_3 = 0$ and $\nu_3 = 1$, the transitions are of very similar wavenumbers.

Another case of inversion doubling is in the first excited singlet state of formaldehyde, shown in section 6.6 to be $^1A_2$ assuming a planar configuration. However, although formaldehyde is planar in the ground electronic state it is pyramidal in the first excited state; but the barrier to inversion is of the intermediate type and all the energy levels are classified according to the $\mathbf{C_{2v}}$ point group. As planar formaldehyde (figure 7.24(a)) becomes non-planar (figure 7.24(b)) it is the $\sigma_v(xz)$

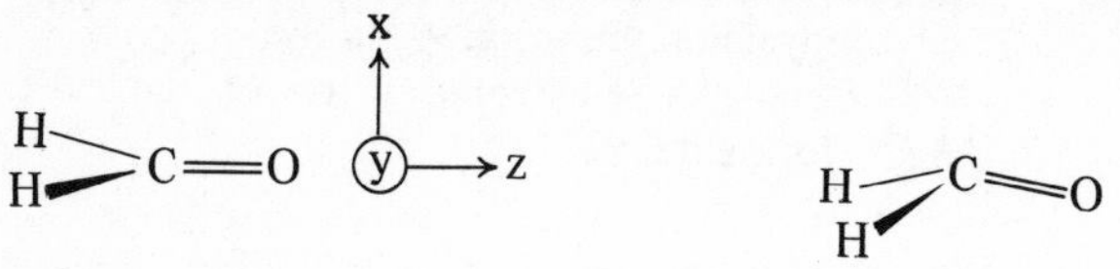

(a) (b)

FIGURE 7.24
(a) Planar and (b) non-planar formaldehyde

plane which is preserved in going from the $\mathbf{C_{2v}}$ to the $\mathbf{C_s}$ point group, the point group which would be relevant if the barrier to inversion were high. Thus the symmetry species in the two point groups should be correlated as in table 7.9. This shows that the $^1A_2$ excited state would be $^1A''$ in

TABLE 7.9
Correlation of symmetry species in the $\mathbf{C_s}$ and $\mathbf{C_{2v}}$ point groups

$\mathbf{C_s}$	σ
A'	1
A''	−1

$\mathbf{C_{2v}}$	$\sigma_v(xz)$	$\sigma_v(yz)$
A_1	1	1
A_2	−1	−1
B_1	1	−1
B_2	−1	1

the C_s point group. When the barrier to inversion becomes of the intermediate type the A″ level splits according to table 7.9 into an A_2 and a B_2 level. In the planar ground state the ground electronic level which would have been A′ in the C_s point group splits into A_1 and B_1 where the B_1 level involves one quantum of the inversion vibration ν_4. As shown in figure 7.25

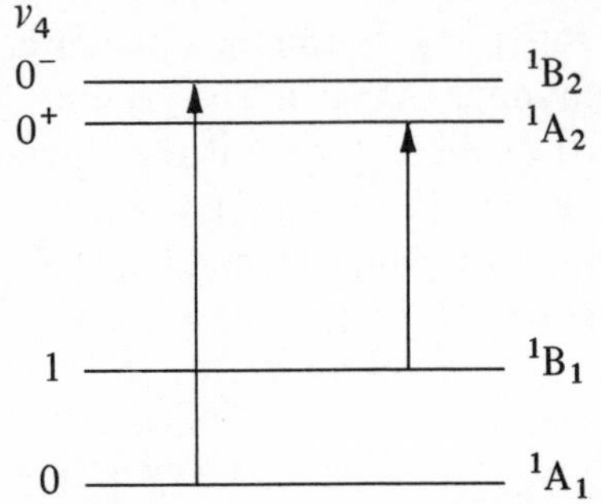

FIGURE 7.25
Vibronic transitions, involving the inversion vibration ν_4, in the $^1A_2 - {}^1A_1$ electronic system of formaldehyde

there are two electric dipole allowed transitions $^1B_2 - {}^1A_1$ and $^1A_2 - {}^1B_1$ both of which are vibronic transitions and polarized along the y-axis.

Aniline (figure 2.2(c)) is non-planar in the ground state but with a fairly low barrier to inversion of the amino-hydrogen atoms so that the potential energy curve for the inversion vibration is similar to that in the A_2 excited state of formaldehyde and energy levels are classified according to the C_{2v} point group.

As we proceed in figure 7.20 from the intermediate to the zero barrier case we are going from a situation in which we would call the molecule non-planar to one in which we would call it planar. This raises the question ‘When do we call a molecule planar or non-planar?’ The answer to this is largely a matter of definition and one useful definition is that if the $v = 0$ level is below the barrier, as in figure 7.26(a), the molecule is said to be *non-planar* and if it is above the barrier, as in figure 7.26(b) it is said to be *planar.* Sometimes the word *quasi-planar* is used to describe the situation in figure 7.26(b) as opposed to the situation where there is no barrier at all. It should be noted that staggering of the vibrational levels occurs even in the quasi-planar case.

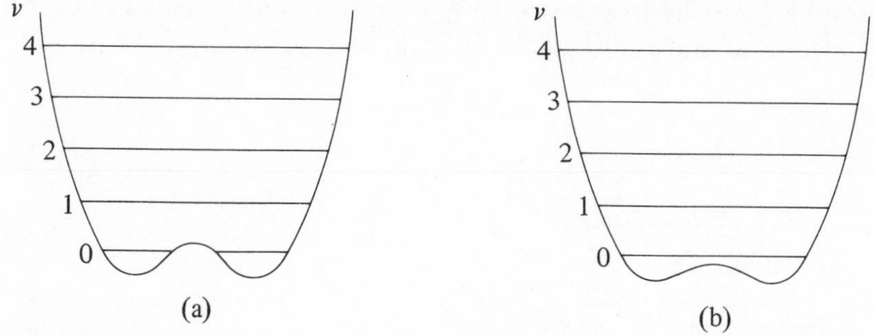

FIGURE 7.26
Potential energy curves for (a) a molecule described as non-planar and (b) a molecule described as planar or quasiplanar

7.8.1(b) Hydrogen-Bonding

Hydrogen-bonding is the process in which a hydrogen atom which is formally bonded to one atom may form a weak bond with a neighbouring atom. The second atom may be in the same molecule (intramolecular hydrogen-bonding) or in another molecule (intermolecular hydrogen-bonding). In either case the situation may be symmetrical, X – H . . . X which is energetically equivalent to X . . . H – X as in the case of the HF_2^- ion, or unsymmetrical, X – H . . . Y which is not equivalent to X . . . H – Y as in intermolecular hydrogen-bonding in phenol and pyridine in which the hydrogen atom of the hydroxyl group of phenol is hydrogen-bonded to the nitrogen atom of pyridine.

In the case of symmetrical hydrogen-bonding the potential energy curve for the X – H stretching vibration is similar to that in figure 7.20(b) except that the superscript + and – now refer respectively to symmetry or antisymmetry with respect to a plane perpendicular to the X . . . X line and bisecting it. The two identical minima in the potential energy curve correspond to the hydrogen atom being more strongly attached to one or the other X.

In the case of unsymmetrical hydrogen-bonding there are still two minima in the potential energy curve for the hydrogen-stretching vibration but they are unequal, as illustrated in figure 7.27. Tunnelling through the barrier can occur near the top but no splitting of levels results because the potential energy curve is unsymmetrical. For example if X and Y are coplanar, as are N and O in the pyridine-phenol case, then the whole system belongs to the

C_s point group and the hydrogen-stretching vibration is of species a′. Since the ground electronic state is A′ all the vibronic levels in figure 7.27 are of

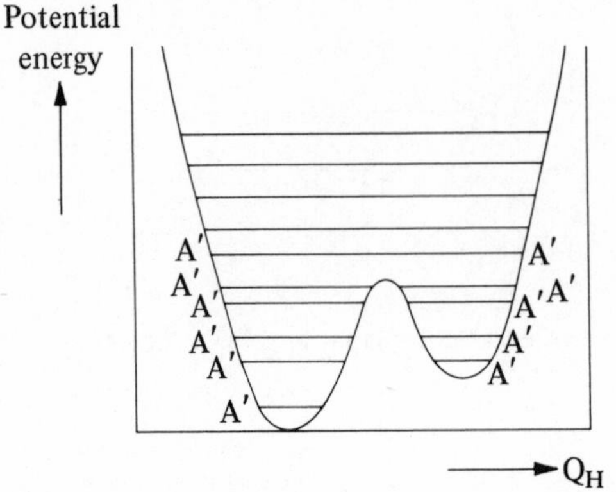

FIGURE 7.27
Potential energy curve for a hydrogen-stretching vibration where the hydrogen atom is involved in unsymmetrical hydrogen-bonding

species A′ and, because of this, interaction can occur by tunnelling through the barrier causing a closing up of the levels in the region of the top of the barrier but no splitting.

7.8.2 POTENTIAL ENERGY CURVES FOR TORSIONAL VIBRATIONS

7.8.2(a) Torsional Vibrations With Unequal Potential Minima

Torsional vibrational motions in molecules with fairly low symmetry such as acrolein, butadiene and glyoxal, illustrated in figure 7.28, may have potential

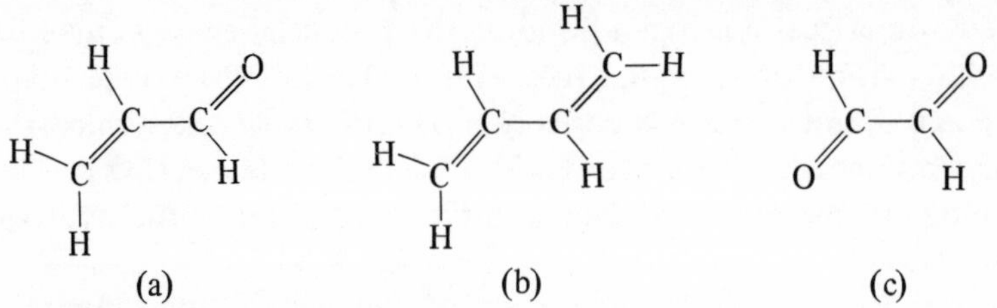

FIGURE 7.28
(a) Acrolein, (b) butadiene and (c) glyoxal

energy curves which show two potential minima corresponding to *s-trans* and *s-cis* forms of the molecule shown for acrolein in figure 7.29. Such a potential energy curve is shown in figure 7.30 in which the normal co-ordinate is approximately the torsional angle. In a case like this with two unequal potential minima there can be no splitting of levels due to tunnelling through

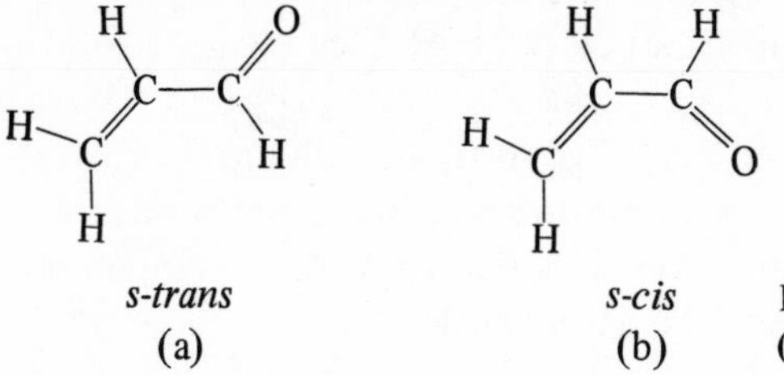

FIGURE 7.29
(a) *s-trans* and (b) *s-cis* isomers of acrolein

the barrier since, as for unsymmetrical hydrogen-bonding, no additional symmetry is introduced by the second minimum. So, if figure 7.30 is the potential energy curve for the ground electronic state of acrolein, which belongs in both *s-cis* and *s-trans* forms to the $\mathbf{C}_s$ point group, the $v = 0$ level

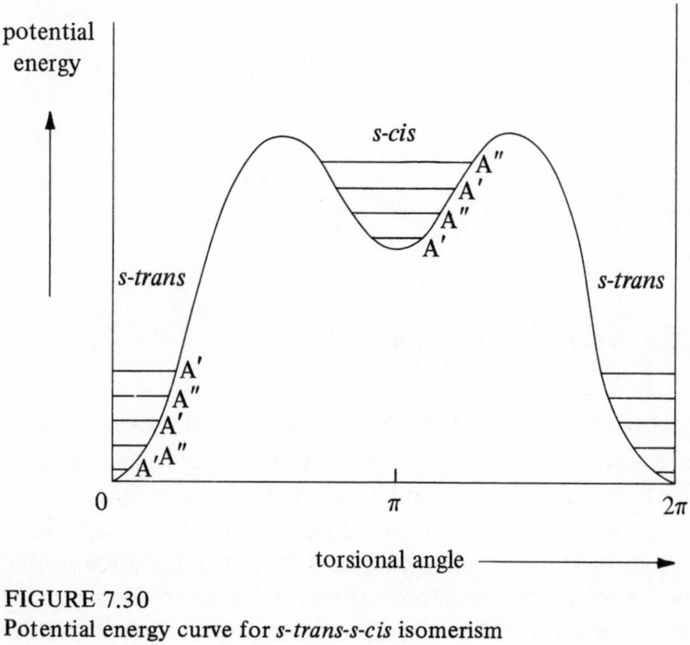

FIGURE 7.30
Potential energy curve for *s-trans-s-cis* isomerism

has A′ symmetry, $\nu = 1$ has A″ symmetry and they alternate thereafter in both potential wells. As the levels approach the top of the barrier they will not be split but levels of the same symmetry will interact through the barrier and disturb the even spacings. Above the barrier rotation about the single bond is free and the energy levels are then those associated with free internal rotation.

Although butadiene and glyoxal belong to the $\mathbf{C_{2h}}$ point group in the *s-trans* form and to the $\mathbf{C_{2v}}$ point group in the *s-cis* form only one element of symmetry, the plane of the molecule, is common to both isomers and the torsional vibrational energy levels should be classified as A′ or A″ for these purposes and the situation is then the same as in acrolein.

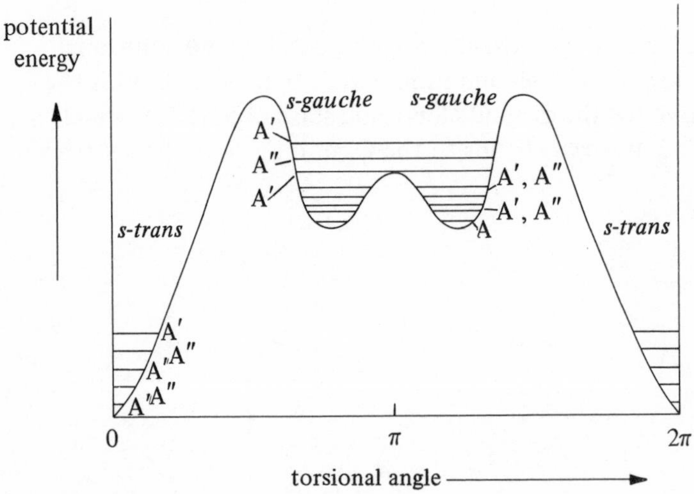

FIGURE 7.31
Potential energy curve for *s-trans-s-gauche* isomerism

It is possible in any of the three molecules in figure 7.28 that it is not an *s-cis* form which corresponds to a second minimum in the potential energy curve but an *s-gauche* form in which the torsional angle is somewhere between 0 and π. A potential curve representing this situation is shown in figure 7.31. In this case tunnelling through the *gauche-gauche* barrier can produce splitting of levels. In the *s-gauche* form acrolein, for example, belongs to the $\mathbf{C_1}$ point group and the unsplit vibrational levels will all be of species A: but when the levels are split by tunnelling they will do so into an A′ and an A″ component which may then interact with the *s-trans* levels towards the top of the *s-trans-s-gauche* barrier causing a very complex pattern of levels.

There can also be torsional vibrations in which there are more than two unequal potential minima. This is the case in ClFHC . OH in which the potential energy curve for torsional motion about the C–O bond shows three unequal minima. The behaviour of the vibrational energy levels will be similar to the two-minima case in figure 7.30.

7.8.2(b) Torsional Vibrations With Equal Potential Minima

In molecules like ethylene ($H_2C = CH_2$), methyl alcohol ($H_3C . OH$) and nitromethane ($H_3C . NO_2$) the potential energy curves shows respectively two, three and six equal potential minima for torsional motion about the C=C, C–O, and C–N bonds. Figure 7.32 shows that there are two

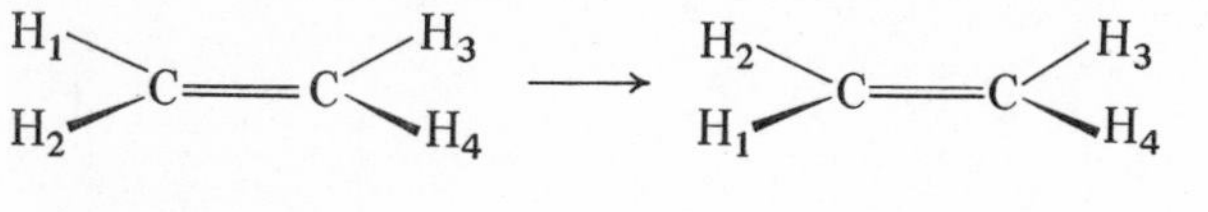

FIGURE 7.32
Two energetically equivalent stable forms of ethylene

energetically equivalent stable forms of ethylene which are interconvertible by a torsional motion but the barrier for ethylene between the two forms is very high. In methyl alcohol there are three energetically equivalent stable forms as illustrated in figure 7.33. The barrier between the forms is low

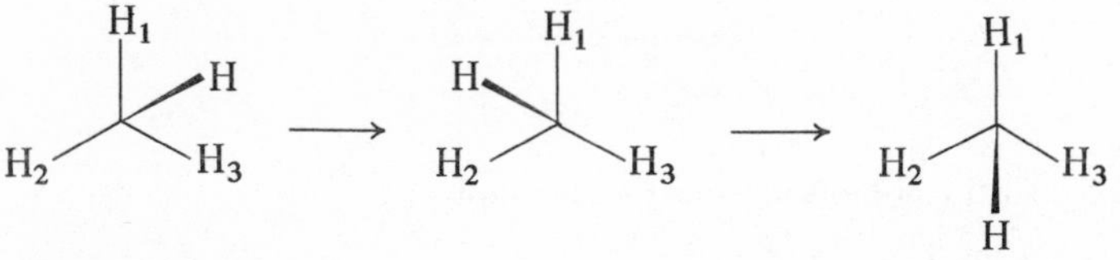

FIGURE 7.33
Three energetically equivalent stable forms of methyl alcohol viewed down the C–O bond

enough to enable splitting to be observed even in the $\nu = 0$ torsional level. For nitromethane figure 7.34 shows that there are six equivalent stable forms, three in which either H_1, H_2, or H_3 is in the position labelled 1 and three in

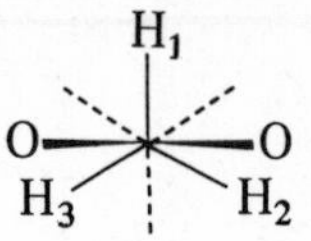

FIGURE 7.34
This figure shows that there are six energetically equivalent stable forms of nitromethene. The molecule is viewed down the C–N bond

which all the hydrogen atoms are rotated through $\pi/3$ as shown by the dotted lines.

The torsion vibrational energy levels in ethylene, methyl alcohol and nitromethane are split respectively into two, three and six components when tunnelling through the barrier is appreciable. In ethylene the splitting is unobservably small. In methyl alcohol the splitting is into three components, two of which are degenerate and alternatively above and below the non-degenerate one, and is 1·3 cm^{-1} for the $v = 0$ level. Figure 7.35 illustrates the potential

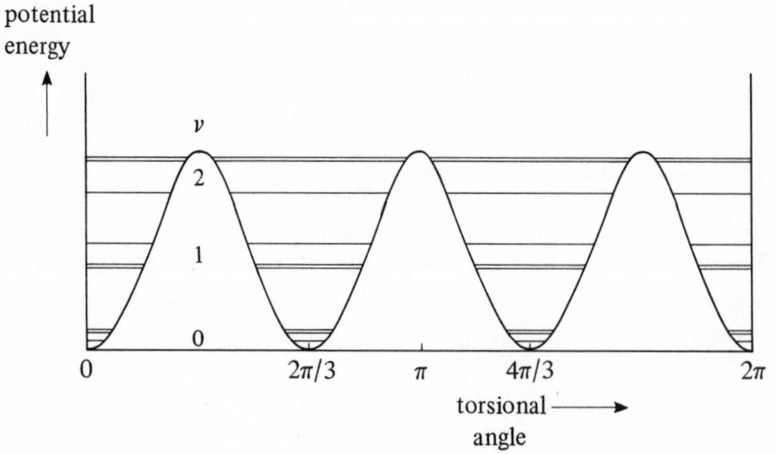

FIGURE 7.35
Potential energy curve for torsional motion in methyl alcohol

energy curve for such a case. In nitromethane the barrier between the potential minima is so low that there is almost free rotation about the C–N bond. This is typical of a situation in which several potential minima occur: toluene ($C_6H_5 . CH_3$) also has six minima in the potential energy curve and almost free rotation about the C–CH_3 bond.

As in the case of the inversion of ammonia, if vibrational energy levels are split by tunnelling through a barrier between two equal minima, the levels should be classified according to the symmetry species of a group of higher order than that to which the equilibrium nuclear configuration belongs. In the case of ammonia it was easy to see that the higher order group is $\mathbf{D}_{3h}$ but for the examples of ethylene, methyl alcohol and nitromethane we cannot describe the higher order group in terms of the symmetry elements described in chapter 2. Additional symmetry elements have to be introduced and the reader is referred to a paper by Longuet-Higgins (Molecular Physics, **6**, 445 (1963)) for an introduction to higher order groups.

7.9 Some limitations of the applications of molecular symmetry

We have seen in the applications of molecular symmetry discussed both in chapter 5 and in this present chapter that molecular symmetry arguments can give us 'yes' or 'no' answers to specific questions; for example 'Are these two protons equivalent?' 'Does this molecule have a dipole moment?' 'Is this molecule optically active?' 'Will this reaction proceed thermally in a conrotatory or disrotatory mode?' 'Is this transition allowed or forbidden?' But there are many cases where protons are *nearly* equivalent, molecules have *almost* zero dipole moment, optical activity is *weak*, and so on. For example in $CH_2{}^{35}Cl^{37}Cl$, compared to $CH_2{}^{35}Cl_2$, the ^{37}Cl atom is a very small perturbation in respect of the vibrational spectrum. We could say that although $CH_2{}^{35}Cl^{37}Cl$ belongs to the $\mathbf{C}_s$ point group it behaves vibrationally almost as if it belongs to the $\mathbf{C}_{2v}$ point group.

We have seen in section 7.1 that the Woodward-Hoffmann rules apply strictly in the two reactions discussed, concerted cycloaddition of two ethylene molecules and concerted isomerization of cyclobutene, only if the hydrocarbons are unsubstituted. If some of the hydrogen atoms in the hydrocarbons are substituted, the strict symmetry arguments no longer apply: but in the cycloaddition of two monofluoroethylene molecules, for example, the fluorine atom perturbs ethylene sufficiently weakly that, if it is assumed that monofluoroethylene belongs to the $\mathbf{D}_{2h}$ point group the correct conclusions will be reached.

In conclusion, then, it must be stressed that apparently rigid symmetry arguments can and should be relaxed in situations which demand it.

Index

Names of molecules. It is the policy in this index to arrange alphabetically the *names* of molecules for which these are well known or easily deduced: for example benzene (C_6H_6) is under 'B' but not under 'C'. Where no simple name is available arrangement is alphabetical according to the *formula*: for example the NH_2 radical is under 'N'.

E

F